Forestry Sciences

Volume 88

Series Editor

Manuel Esteban Lucas Borja
Dept. of Agroforestry Technology and Science and Genetics
Castilla La Mancha University
Albacete, Spain

The aim of the book series *Forestry Sciences* is to present state-of-the-art results on the research, development, and application of forest sciences. Broad in scope, the series covers nearly all aspects related to the natural sciences of forests, including but not limited to forest soils, forest hydrology, tree genetics and breeding, tree biotechnology, forest biometrics, modeling and mapping, and forest fire and fuel management. Volumes on methodology with a theoretical emphasis are especially welcome. The series publishes solicited or unsolicited Contributed Volumes (a collection of reviews written by experts) and Monographs (a comprehensive overview of a specific topic written by one or multiple authors) with high academic quality. This collection appeals to a broad audience, including international scientists, advanced students, forest managers, and practitioners.

Sheriza Mohd Razali • Zaiton Samdin
Norfaryanti Kamaruddin
Ruzana Adibah Mohd Sanusi
Editors

Sustainable Eucalyptus Forestry in Southeast Asia

Balancing Socio-Economic Benefits and Environmental Impact with Innovative Technologies

Springer

Editors
Sheriza Mohd Razali
Institute of Tropical Forestry and Forest Products
Universiti Putra Malaysia
Serdang, Malaysia

Norfaryanti Kamaruddin
Institute of Tropical Forestry and Forest Products
Universiti Putra Malaysia
Serdang, Malaysia

Zaiton Samdin
School of Business and Economics
Universiti Putra Malaysia
Serdang, Malaysia

Ruzana Adibah Mohd Sanusi
Faculty of Forestry and Environment
Universiti Putra Malaysia
Serdang, Malaysia

ISSN 0924-5480 ISSN 1875-1334 (electronic)
Forestry Sciences
ISBN 978-3-032-21342-6 ISBN 978-3-032-21343-3 (eBook)
https://doi.org/10.1007/978-3-032-21343-3

This Springer imprint is published by the registered company Springer Nature Switzerland AG
The registered company address is: Gewerbestrasse 11, 6330 Cham, Switzerland

Preface

The raw material shortage in the wood-based industry has been a long-winded issue. Various solutions have been planned and implemented, including establishing large-scale forest plantations in Malaysia. The establishment is also intended to release the pressure on natural forest resources. As of March 2024, the Sabah Forestry Department reported ongoing efforts to establish up to 400,000 hectares of forest plantations by 2036, with an annual planting rate target of 18,000 hectares per year (Borneo Post Online, 2024). Sarawak has established approximately 594,553 hectares of industrial plantations and aims to achieve 1 million hectares of planted forest by 2025; however, only about 56% of the Licensed Planted Forest (LPF) zones are deemed suitable for planting.

There are 11 species initially approved for the Forest Plantation Development Program (PPLH): Bambuseae (Buluh), *Tectona grandis* (Jati), *Azadirachta excelsa* (Sentang), *Khaya ivorensis* (Khaya sp.), *Neolamarckia cadamba* (Kelempayan/Laran), *Paraserianthes falcatari* (Batai), *Octomeles sumatrana* (Binuang), Rubberwood (Timber Latex Clone), *Acacia* spp. (mangium/hybrid), *Paulownia* spp., and *Eucalyptus* spp. Malaysian Timber Industrial Board (MTIB) is the agency responsible for disbursing the fund allocated for the forest plantation program. The program was approved in 2006 with an allocation of RM 1.045 billion to achieve 375,000 hectares of forest plantation in Peninsular Malaysia.

Various stakeholders are involved in governing and supporting roles in the program. Both the federal and state government agencies are assuming active roles to ensure the success of the program. Among the key agencies are the Forestry Department of Peninsular Malaysia (JPSM), Ministry of Natural Resources and Environmental Sustainability (NRES), Sarawak Forest Department, Sabah Forest Department, Malaysia Rubber Board (LGM), Sabah Forestry Development Authority (SAFODA), Sarawak Timber Industry Development Corporation (STIDC), and Forest Research Institute Malaysia (FRIM). Some of the challenges that may be faced by stakeholders in collaborative governance aimed at improving forest management, as observed in Sabah, Malaysia, are stakeholder representation, accountability, and goal alignment. These challenges may impact the planning and monitoring of forest plantation initiatives.

Moreover, there are conflicting views on *Eucalyptus* plantation, especially on the ecological and social impact, in addition to the lack of current information on silviculture and management, pests and diseases, productivity, and economic and market aspects, especially from the industrial experiences and perspectives. This creates skepticism and uncertainties among various stakeholders, especially potential investors and financial institutions that provide financial assistance. Thus, it is crucial to understand the current affairs in *Eucalyptus* plantations and wood-related industries' condition and stance in order to illuminate the conflicting issues on *Eucalyptus* plantations. Updated information on silviculture and management, pests and diseases, productivity, and socioeconomic as well as industrial perceptions related to *Eucalyptus* plantations would be useful to plantation owners, potential investors, wood manufacturing companies, and related government agencies and universities. This would stimulate the plantation, which can then promote the sustainability of Malaysia's wood-based industry and hence the socioeconomic growth of the local community.

This book aims to highlight and discuss the research findings, specifically those related to *Eucalyptus* forest plantations in Malaysia and other tropical countries such as Indonesia. Following this introductory chapter, there are 11 chapters that discuss various topics ranging from upstream to downstream of the *Eucalyptus* plantation. All these chapters are classified into three main thematic sections, focusing on ecology, forest management, and socio and circular economics.

Ecology Perspectives

The first section presents the scope of the Ecology perspectives, highlighting upstream topics. In this section, topics such as micropropagation, tech-driven agroforestry soil conservation, microclimate, and forest fire as well as wildlife and ecological response in relation to *Eucalyptus* plants and plantations are discussed. In the first chapter of this section, *Eucalyptus* micropropagation using in vitro approach is presented. It reviews the advancement in the micropropagation technique that can be used to counter challenges in *Eucalyptus* propagation such as low germination and genetic variability. The chapter also reviews previous micropropagation studies on *Eucalyptus* species to pique the interest in alternate sources of high-quality *Eucalyptus* planting materials and their prospects. It also highlights biotechnology as a solution to produce genetically uniform and disease-free seedlings in supporting the growing demand for superior wood quality in forest plantations.

The second chapter discusses tech-driven agroforestry soil conservation, where agroforestry systems that integrate trees, crops, and livestock can be utilized as sustainable solutions for soil degradation. This research focuses on case studies across tropical and semiarid regions in the mitigation of soil degradation within agroforestry landscapes through technologies for climate resilience, biodiversity, and rural development. It highlights the utilization of technologies such as Geographic Information System (GIS), Internet of Things (IoT), Artificial

Intelligence (AI), and bioinformatics as parts of Agroforestry 2.0, which allows precision soil health monitoring, nutrient cycling, and resilient land-use planning.

The third chapter covers the topic of microclimate and forest fire risk in *Eucalyptus* plantation. This chapter discusses *Eucalyptus* forest plantation functions as provider of ecosystem services to the surrounding environment through the modification of the microclimate conditions but also has high susceptibility to fire. This review explores the link between *Eucalyptus* plantations, climate conditions, and fire incidence by synthesizing previous studies to highlight current knowledge and research gaps. It also highlights the vulnerability and fire-related risks associated with the species, thus addressing the need for further studies and improved management strategies to mitigate fire risks associated with plantation design and species selection.

The fourth chapter in the Ecology section highlights the wildlife in *Eucalyptus* forest plantations and the issue of the planted forests' limited ability to support all wildlife species, as natural forests do. This chapter reviews studies on wildlife biodiversity in *Eucalyptus* plantations and discusses their benefits and explores the factors influencing the survival of certain species. It offers recommendations to enhance biodiversity through the improvement of planting practices and the integration of conservation into plantation management.

Forest Management Perspectives

The second section presents the scope of the Forest Management perspectives, highlighting research in enhancing productivity, wood quality, and environmental monitoring. In this section, three topics such as innovative wood quality assessment, forest fire detection technology, and interaction between *Eucalyptus* and tropical mangrove ecosystems are discussed, enabling better management of *Eucalyptus* forest plantation. The first chapter of this section explores the application of near-infrared spectroscopy (NIR) in measuring wood quality of plantation-grown *Eucalyptus* trees. This chapter reviews the use of NIR as an efficient tool as the application is nondestructive and cost-effective in evaluating wood quality traits in terms of chemical, physical, anatomical, and mechanical properties. It further emphasizes the need for technological advancements in forest management, specifically in assessing the impact of silvicultural practices on wood quality.

In the second chapter of this section, it covers the utilization of satellite-based remote sensing technology in forest fire detection, highlighting its capability to assist in enhancing community safety and environmental monitoring. This chapter highlights the role of remote technologies in *Eucalyptus* plantation to detect, track, and analyze fire events that support real-time risk assessment and early warning systems, improving human safety and ecological protection as well as presenting case studies from other countries. It further emphasizes the importance of post-fire assessments and government incentives, in addition to remote sensing applications in *Eucalyptus* forest risk management and planning.

Interestingly, the third chapter of this section represents the challenges and opportunities in tropical Mangrove plantations and their ecological interaction with *Eucalyptus* plantations. This chapter explores large-scale mangrove restoration initiatives in Indonesia, alongside addressing the ecological, economic, and policy challenges. It also highlights the interactions between mangrove and *Eucalyptus* plantations, particularly in regions where land conversion or buffer planting affects hydrological and ecological dynamics. This chapter recommends integrated approaches in forest management and planning, specifically to balance wood production with water resource conservation to avoid negative ecosystem impacts.

Socio and Circular Economic Perspectives

The third section of the book consists of three chapters that look into *Eucalyptus* in supporting the circular economy and socioeconomic development. It covers topics such as the use of *Eucalyptus urophylla* for natural fabric color binding, *Eucalyptus* as a source of bioproducts, and circular economy and socioeconomic impacts of *Eucalyptus* plantations. The first chapter explores the utilization of *E. urophylla* as a source of tannin for natural color binding on fabrics using the eco-printing technique for supporting circular economy. It showcases the environmentally friendly quality of this approach compared to traditional approaches due to its minimal use of chemicals and supports waste reduction. The study highlights the potential of *E. urophylla* in producing tannins and other beneficial phytochemicals for improved color binding, retention, and leaching resistance without altering fabric texture.

For the second chapter, it presents the potential of *Eucalyptus* to produce a variety of valuable bioproducts as well as discusses the circular economy impact on the wood industries and society. It covers the importance in utilizing *Eucalyptus* biomass to reduce waste and generate new economic opportunities for smallholders, farmers, and entrepreneurs. The chapter includes the chemical composition of *Eucalyptus* and its industrial applications, emphasizing how it supports circular economy goals by transforming biomass into high value products.

The third chapter of this section reviews the socioeconomic impact of the *Eucalyptus* plantation that emphasizes the contribution to source of income to the local community, employment, and infrastructure development. It explores the role of *Eucalyptus* plantations to improve the socioeconomics of society and provide the local community with better livelihood including increase in job opportunities. This chapter adds on how *Eucalyptus* plantations contribute to socioeconomic development by providing employment, income generation, and infrastructure development. It also covers information on potential products and services from *Eucalyptus*, food security considerations, natural resources management, and sustainability as well as addressing the ongoing uncertainties on social impact of *Eucalyptus* plantations.

This book ends with a final chapter that provides an overall conclusion that specifically highlights the future direction of *Eucalyptus* plantations. This book presents a comprehensive discussion on the key aspects of the *Eucalyptus* plantations providing insights to researchers, students, and other stakeholders into the potential of sustainable *Eucalyptus* plantations through integrating the socioeconomic benefits and environmental impact with innovative technologies.

Serdang, Malaysia

Sheriza Mohd Razali
Zaiton Samdin
Norfaryanti Kamaruddin
Ruzana Adibah Mohd Sanusi

Contents

Socioeconomic and Circular Economy Perspectives

About the Editors

Sheriza Mohd Razali is a prominent researcher with a strong background in the field of remote sensing, environmental science, and tropical forest management. She has made significant contributions to the study of forest ecosystems, biodiversity conservation, and environmental monitoring, particularly in tropical regions. Her research expertise spans across various areas, including the assessment of forest vegetation health, land cover classification, and climate change impacts. She has worked extensively with satellite imagery, utilizing remote sensing techniques to monitor and analyze forests, plantations, and agricultural areas. One of her key areas of focus is the use of remote sensing indices like the Normalized Difference Vegetation Index (NDVI) and the Normalized Difference Water Index (NDWI) for environmental monitoring. These techniques have been crucial in her studies on forest health, peatland fire management, and vegetation productivity. Throughout her career, she has authored numerous articles and publications, many of which have been featured in highly regarded international journals such as *Fire*, *Forests*, *Pertanika Journal of Science and Technology*, and *Journal of Tropical Biodiversity and Biotechnology*. Her work is recognized for its impact on improving land and forest management practices, as well as its applications in climate change mitigation and sustainable development.

Zaiton Samdin is a Professor at the School of Business and Economics and the Deputy Director at the Institute of Tropical Forestry and Forest Products (INTROP) at Universiti Putra Malaysia. She holds a PhD from the University of Exeter, United Kingdom. Her research interests lie in economic valuation and tourism economics, with a strong focus on sustainable tourism development. She explores a range of issues including the socioeconomic and environmental impacts of forest plantation establishment, the public's valuation of ecosystem services (such as endangered species, national parks, and protected areas), and how these valuations can influence policymaking and decision-making and promote behavior change. With over a decade of expertise in valuation economics, she has led numerous capacity-building projects aimed at enhancing the decision-making abilities of government agencies, nongovernmental organizations, and natural resource managers. These initiatives emphasize the use of high-quality evidence and analytical methods to support better decisions for sustainable development.

Norfaryanti Kamaruddin is a researcher from Malaysia with a strong background in environmental sustainability, supply chain management, and education. She has contributed extensively to research in areas such as big data analytics, sustainable supply chains, environmental impacts of agricultural practices, and decision-making within industries like palm oil and forestry. Throughout her career, she has collaborated with other experts on a variety of interdisciplinary studies and has authored or co-authored several influential papers in reputable journals.

Ruzana Adibah Mohd Sanusi is an accomplished researcher and academic known for her significant contributions to the fields of urban ecology, environmental science, and sustainable development. Her work primarily focuses on the relationship between urban environments, biodiversity, and micro-climates, with a particular interest in the role of vegetation in mitigating climate change impacts. Her academic journey has seen her engage in a wide range of research that includes studies on the ecological dynamics of urban forests, agroforestry systems, and the impact of green infrastructure on urban cooling and human thermal comfort. She has been involved in exploring how different land use systems—such as oil palm and rubber plantations—affect biodiversity and microclimates, especially in tropical regions.

Ecological Perspectives

Micropropagation of *Eucalyptus* Species

Haida Zainol, Ruzana Adibah Mohd Sanusi, and Hakiman Mansor

Abstract *Eucalyptus* is an economically significant plantation tree belonging to the Myrtaceae family and comprises more than 700 species worldwide. This species is valued for its rapid growth and commercial importance in timber, pulp, and bioenergy production, making it a widely cultivated resource. *Eucalyptus* is conventionally propagated by seeds; however, low germination rates, seed dormancy, and genetic variability affect plantation quality and productivity. To overcome these limitations, tissue culture has emerged as an effective solution for mass propagation in *Eucalyptus*. Micropropagation enables the mass production of genetically uniform, disease-free *Eucalyptus* seedlings with desirable traits. This chapter focuses on recent advancements in *Eucalyptus* micropropagation, specifically examining studies published between 2018 and 2024. The key factors influencing successful micropropagation, including the optimization of basal medium formulation and the application of plant growth regulators, will be synthesized and discussed.

H. Zainol
Institute of Tropical Forestry and Forest Products, Universiti Putra Malaysia,
Serdang, Selangor, Malaysia

R. A. Mohd Sanusi (✉)
Institute of Tropical Forestry and Forest Products, Universiti Putra Malaysia,
Serdang, Selangor, Malaysia

Department of Forestry Science and Biodiversity, Faculty of Forestry and Environment,
Universiti Putra Malaysia, Serdang, Selangor, Malaysia
e-mail: ruzanasanusi@upm.edu.my

H. Mansor
Institute of Tropical Forestry and Forest Products, Universiti Putra Malaysia,
Serdang, Selangor, Malaysia

Department of Crop Science, Faculty of Agriculture, Universiti Putra Malaysia,
Serdang, Selangor, Malaysia

S. Mohd Razali et al. (eds.), *Sustainable Eucalyptus Forestry in Southeast Asia*,
Forestry Sciences 88, https://doi.org/10.1007/978-3-032-21343-3_1

1 Introduction

Eucalyptus is a fast-growing, evergreen genus that belongs to the Myrtaceae family. It is indigenous to Australia, with more than 900 species also found in Brazil, Chile, Argentina, India, France, Indonesia, Morocco, Portugal, Spain, South Africa, and the United States, covering approximately 20 million hectares globally (Trueman et al. 2018; Sharma et al. 2023). *Eucalyptus* is a highly adaptable species that can thrive in a wide array of environments, ranging from arid to temperate climates. In addition to being able to tolerate extreme abiotic environments such as drought and poor soil conditions, *Eucalyptus* is also resilient to various biotic stressors, even exhibiting high productivity. These advantageous traits make it a valuable resource with high market demand in the pulp and hardwood industries (Singh et al. 2021; Qin et al. 2022; Zhu et al. 2022). In addition, *Eucalyptus* produces valuable essential oils extracted from various parts of the plant that possess numerous biological properties with promising prospects in the pharmaceutical and nutraceutical industries (Almas et al. 2021; Čmiková et al. 2023; Khedhri et al. 2024). Due to its diverse applications across numerous industries, the demand for *Eucalyptus* has steadily increased, highlighting the need for efficient and rapid propagation methods for large-scale production of *Eucalyptus* raw materials.

Seed propagation is the most common way of cultivating *Eucalyptus*, but it has notable drawbacks, including poor genetic stability and long breeding cycles (Wang et al. 2024). To overcome these limitations, vegetative propagation techniques such as layering, grafting, and cuttings have been employed for the asexual reproduction of *Eucalyptus*. These techniques serve as alternatives to seed propagation and are effective in increasing the production of *Eucalyptus* seedlings. However, their success largely depends on the performance and physiological condition of the donor tree (Vahdati et al. 2022). Although conventional propagation techniques have been the primary methods for forest plantations, recent technological advancements have led to the emergence of plant tissue culture as a promising technique for the mass production of *Eucalyptus* seedlings.

The in vitro approach has been widely adopted for propagating various plant species, including *Eucalyptus* (Faria et al. 2021; de Oliveira et al. 2022). In addition to commercial applications, plant tissue culture ensures the continuous production of plant species, particularly those that are endangered or highly utilized. Plant tissue culture is defined as the technique of growing any part of a plant, including cells, tissues, or organs, on artificial nutrient medium under controlled environmental conditions (Mehbub et al. 2022). In addition to mass production, plant tissue culture has become essential for producing disease-free plantlets, improving genetic traits, preserving germplasm, and facilitating the production of valuable secondary metabolites (Fazili et al. 2022).

2 Plant Tissue Culture Components

Plant tissue culture involves growing plant cells, tissues, or organs under controlled laboratory conditions. Its success depends on several key components, each of which plays a crucial role in providing the optimal conditions for plant growth and development (Bidabadi and Jain 2020). These components can be categorized into three main areas: the basal medium composition, the selection and preparation of planting materials, and the control of environmental factors.

2.1 Basal Medium Composition

In plant tissue culture, the basal medium is the foundation for the growth and development of plant cells, tissues, or organs. It contains essential nutrients to support cell division and differentiation in vitro. The composition of the basal medium is critical for successful propagation, as it provides all the basic elements required by plants for cellular metabolism, division, and regeneration (Mehbub et al. 2022). The key components of the basal medium included macronutrients, micronutrients, vitamins, amino acids, carbohydrates, plant growth regulators, gelling agents, and pH buffers.

2.1.1 Macronutrients

Macronutrients are major inorganic elements required by plants in relatively large amounts, typically at concentrations above 0.5 mM (Hameg et al. 2020). These include nitrogen (N), phosphorus (P), potassium (K), magnesium (Mg), calcium (Ca), and sulfur (S). Carbon (C), hydrogen (H), and oxygen (O) are also considered macronutrients but are supplied in gaseous form.

- *Nitrogen (N)*: Supplied as ammonium nitrate (NH_4NO_3) or potassium nitrate (KNO_3), which are essential for growth and protein synthesis.
- *Phosphorus (P)*: Provides potassium phosphate (KH_2PO_4) or calcium phosphate ($Ca_3(PO_4)_2$), which are crucial for cell division and energy transfer.
- *Potassium (K)*: This potassium is supplied as potassium nitrate (KNO_3), which aids in osmotic regulation and enzyme activation.
- *Calcium (Ca)*: $CaCl_2$ is important for cell wall stability and membrane function.
- *Magnesium (Mg)*: Supplied as the magnesium sulfate ($MgSO_4$) necessary for chlorophyll synthesis and enzyme activation.
- *Sulfur (S)*: Sulfur is supplied as magnesium sulfate ($MgSO_4$) or sodium sulfate (Na_2SO_4), which are essential for the synthesis of proteins and vitamins.

2.1.2 Micronutrients

Micronutrients are trace elements that are required in small quantities (less than 0.05 mM) but are essential for supporting enzyme functions, growth regulation, and metabolic processes (Hameg et al. 2020). These micronutrients are as follows:

- *Iron (Fe)*: Crucial for chlorophyll production and electron transport.
- *Manganese (Mn), zinc (Zn), copper (Cu), molybdenum (Mo), boron (B), and cobalt (Co)*: Necessary for enzymatic activity, nutrient uptake, and cell division.

2.1.3 Vitamins

Vitamins are organic compounds supplemented in the basal medium that regulate cellular processes. The commonly used vitamins are as follows:

- *Thiamine (vitamin B1)*: Involved in carbohydrate metabolism and energy production.
- *Pyridoxine (vitamin B6)*: Essential for amino acid metabolism.
- *Nicotinic acid (vitamin B3)*: Involved in energy production and stress tolerance.
- *Myo-inositol*: Important for phospholipid synthesis and growth regulation.

2.1.4 Amino Acids

Amino acids are organic compounds that serve as building blocks for protein synthesis and growth. They are important in promoting cell division and differentiation.

2.1.5 Carbohydrates

Carbohydrates, such as sucrose, are supplied primarily in sugar form and are essential components of plant tissue culture basal medium. In their natural habitat, plants are autotrophs that produce their own food through photosynthesis via carbon dioxide, light, water, and other elements obtained from their surroundings. However, under in vitro conditions, plant cells or plantlets are grown under heterotrophic or mixotrophic conditions. Thus, sugar is supplemented in the basal medium as a source of energy to provide carbon for the plant to conduct physiological processes (Isah and Umar 2020). In addition, sugar is added to control the osmotic pressure of the basal medium (Sohrabi et al. 2021). In the micropropagation of *Eucalyptus* species, sucrose at concentrations of 2 to 3% has been found to promote optimal growth during the shoot initiation, shoot proliferation, and rooting stages (Frade et al. 2023; Karet et al. 2021; Avelar et al. 2024).

2.1.6 Plant Growth Regulators

Plant growth regulators are organic or synthetic compounds with low molecular weights that influence various physiological and metabolic processes in plants, including shoot and root induction, seed germination, and callogenesis (Karakas 2020; Cavallaro et al. 2022). They can be classified as natural or synthetic plant hormones. In plant tissue culture, plant growth regulators are applied exogenously in the basal medium at specific concentrations tailored to achieve specific developmental outcomes. Among the various classes of plant growth regulators, cytokinins and auxins are the most frequently used in plant tissue culture.

Cytokinins are plant growth regulators derived from the nitrogen-containing compound adenine (N^6) and can be divided into adenine-based cytokinins, such as kinetin, zeatin, and 6-benzylaminopurine (BAP), and phenylurea-based cytokinins, such as thidiazuron (TDZ) (Vylíčilová et al. 2020; Brunoni et al. 2021). These regulators are essential for cytokinesis and cell differentiation in shoots and roots (Terceros et al. 2020; Yang et al. 2021). Since isolated explants lack the ability to synthesize adequate amounts of cytokinins, an optimal concentration must be supplied in the basal medium to support regeneration and growth. Auxins represent another class of plant growth regulators that are primarily used in plant tissue culture for root and callus induction (Bano et al. 2022). Their application helps regulate various plant activities, such as cell division and elongation (Gomes and Scortecci 2021). For root induction, supplementation with indole-3-butyric acid (IBA), indole-3-acetic acid (IAA), and 1-naphthaleneacetic acid (NAA) is commonly preferred (Tien et al. 2020). For the callus culture, 2,4-dichlorophenoxyacetic acid (2,4-D), 4-amino-3,5,6-trichloro-2-pyridinecarboxylic acid (picloram), and 3,6-dichloro-2-methoxybenzoic acid (dicamba) are more prominent for initiating and proliferating callus tissues (Gantait and Mahanta 2021; Ahansal et al. 2022).

2.1.7 Gelling Agents

The gelling agent is an inert element used in plant tissue culture to solidify the basal medium mixture. Several types of gelling agents are available on the market, with agar and gelrite being the most commonly used. Agars are gelatinous substances derived from red algae. In *Eucalyptus* micropropagation, agar is supplemented at a concentration between 4 and 8 g/L (Avelar et al. 2020; Molinari et al. 2021; Shwe and Leung 2020; de Oliveira et al. 2022). Moreover, gelrite is a gellan gum-based gelling agent derived from bacteria. It has been used at concentrations of 2.8 and 3 g/L for the in vitro propagation of *E. grandis* × *E. nitens* and *E. grandis* × *E. urophylla* (Karet et al. 2021; Nazirah et al. 2021).

2.2 Planting Materials

In *Eucalyptus* tissue culture, seeds and nodal segments consisting of axillary buds are the most commonly used explants. Seeds are the simplest explants with a low contamination rate and high initiation rate due to the juvenility of young seedlings (Trueman et al. 2018). However, the downside of using seeds as explants is genetic variability, as seed-derived plants are characterized by diverse traits, which can lead to inconsistent growth and development patterns. Furthermore, *Eucalyptus* seeds typically exhibit physical dormancy due to their impermeable, hard seed coat (Afroze et al. 2021). To overcome seed dormancy, specific pretreatments, such as scarification, acid scarification, or hot water treatment, are necessary (Kumar et al. 2024). Alternatively, nodal segments offer the benefits of a high regeneration rate and genetic uniformity. However, they are highly susceptible to contamination, primarily because they are typically collected from mature plants that have experienced diverse environmental conditions and are exposed to numerous pathogens. The shoot tips of *Eucalyptus* can be used as explants. They have a high regeneration rate because of their abundance of actively dividing meristematic cells. However, the number of shoot tips per plant is limited, and shoot tips are highly sensitive to sterilization procedures. The delicate meristematic tissues are easily damaged, often leading to death.

3 Micropropagation of *Eucalyptus*

3.1 Establishing Axenic Culture

The establishment of axenic culture is the most crucial part of the plant tissue culture technique. The term "axenic" refers to a culture that is completely free from contaminants such as fungi, bacteria, and other organisms that can contaminate and interfere with the growth and development of the plant (Heck et al. 2021). Since explants are collected from nonsterile environments, a sterilization procedure is needed to eliminate contaminants. Pretreatment is typically carried out with a fungicide by soaking explants for 20–30 minutes. The samples were then thoroughly washed with running tap water supplemented with a commercial detergent for at least 30 minutes. After that, further sterilization was conducted in a laminar air flow cabinet for aseptic handling. The explants are placed in autoclaved jars and immersed in 70% ethanol for 30 to 60 seconds (de Oliveira et al. 2022; Nazirah et al. 2021;

Avelar et al. 2024). After that, the explants are treated with sterilant agents such as sodium hypochlorite (NaOCl) for 15–30 minutes or mercuric chloride ($HgCl_2$) for 2–4 minutes, followed by rinsing with sterile distilled water (Karet et al. 2021; Nazirah et al. 2021). Wetting agents or surfactants such as Tween 20 or Triton X-100 can be added to prevent drying (Ayala et al. 2019; Karet et al. 2021). Among the commonly used sterilizing agents, sodium hypochlorite is preferable to mercuric chloride because it is relatively safe and less toxic for human handling and plant tissue. In contrast, while mercuric chloride is highly effective, its use is limited because of its highly toxic nature to humans and plants, as well as potential environmental hazards (Animasaun et al. 2024). Hence, sodium hypochlorite is a widely preferred choice for sterilization.

3.2 In Vitro Seed Germination

In vitro germination is the process of germinating seeds on a basal medium under sterile laboratory conditions. This method offers several advantages, including low contamination rates and enhanced germination rates. Various studies have explored the optimal conditions for *Eucalyptus* seed germination. For example, de Oliveira et al. (2022) successfully germinated *E. cloeziana* seeds on half-strength Murashige and Skoog (MS) medium supplemented with 15 g/L sucrose and 4 g/L agar. Similarly, *E. nitens* seeds were germinated on full-strength MS medium supplemented with 30 g/L sucrose and 6.5 g/L agar (Ayala et al. 2019). Moreover, *E. camaldulensis* seeds were germinated on a half-strength MS medium supplemented with 30 g/L sucrose and 4 g/L gelzan (Ahad et al. 2018). Compared with other studies in which MS medium and sucrose were used, *E. microcorys* seeds were successfully germinated on medium solidified with distilled water only (Faria et al. 2021).

3.3 In Vitro Shoot Induction and Proliferation

The success of shoot induction and proliferation in *Eucalyptus* micropropagation depends on the ability of axillary buds to break dormancy and subsequently proliferate. Typically, shoot initiation is conducted on full-strength MS medium, as reported for shoot initiation of *E. pilularis, E. globulus, E. grandis* × *E. nitens, E. grandis* × *E. urophylla, E. microcorys, E. grandis, E. camaldulensis, E. cloeziana, E. pilularis,* and *E. tereticornis* (Avelar et al. 2020; Molinari et al. 2021; Souza et al. 2020; Nazirah et al. 2021; Karet et al. 2021; Esquivel et al. 2024; Avelar et al. 2024). A study by Frade et al. (2023) investigated shoot initiation in *E. dunnii* and *E. grandis* × *E. urophylla* via full-strength woody plant medium (WPM). Once shoots are initiated, the next phase involves shoot proliferation to increase the number of shoots. The basal medium is typically supplemented with a combination of cytokinin and auxin. Previous studies have shown that BAP at concentrations ranging from 0.22 to 4.44 μM is the most frequently used for shoot proliferation, often fortified alongside low concentrations of auxins such as NAA, IAA, and IBA (Molinari et al. 2021; Souza et al. 2020; Nazirah et al. 2021; Karet et al. 2021; Frade et al. 2023; Esquivel et al. 2024; Avelar et al. 2024). Shoot proliferation is also enhanced by adjusting macronutrient, micronutrient, and vitamin formulations. For example, Esquivel et al. (2024) achieved successful shoot proliferation in *E. globulus* via a basal medium containing MS macronutrients and WPM micronutrients. Similarly, Karet et al. (2021) reported improved shoot proliferation in *E. grandis* × *E. nitens* via the combination of the macronutrients and micronutrients of MS formulations with the vitamins of BS formulations.

3.4 Formation of Adventitious Root

Adventitious root formation is a crucial step in preparing plantlets for acclimatization and transfer to the field. Root development is typically induced by transferring the plantlets onto a basal medium formulation similar to that used for shoot proliferation. However, the basal medium strength is often reduced to half to improve rooting success and minimize the risk of hyperhydricity or callus formation, which can interfere with root formation. Studies on *E. grandis* × *E. nitens, E. grandis* × *E. urophylla,* and *E. bosistoana* have shown successful root induction via half-strength MS medium (Shwe and Leung 2020; Nazirah et al. 2021; Karet et al. 2021). Moreover, Esquivel et al. (2024) reported that *E. globulus* achieves optimal root initiation with one-third-strength MS medium. Although half-strength basal medium is recommended for rooting, full-strength WPM and MS medium have also been effective in inducing root formation in *E. pilularis, E. grandis* × *E. urophylla,* and *E. camaldulensis* (Ahad et al. 2018; Souza et al. 2020; Frade et al. 2023; Avelar et al. 2024). In addition to macronutrients, micronutrients, and vitamins, sucrose and agar concentrations also directly affect the root initiation of *Eucalyptus* plantlets. Sucrose supplementation at 10 to 20 g/L has been found to induce the formation of roots in *E. pilularis, E. globulus, E. dunnii, E. grandis* × *E. urophylla,* and *E. grandis* × *E. nitens* (Karet et al. 2021; Frade et al. 2023; Esquivel et al. 2024; Avelar et al. 2024). For the gelling agent, agar at concentrations ranging from 5.5 to 8 g/L is typically used to solidify the basal medium (Shwe and Leung 2020; Souza et al. 2020; Frade et al. 2023; Esquivel et al. 2024; Avelar et al. 2024).

3.5 Acclimatization

Acclimatization refers to the process of gradually adapting tissue-cultured plantlets to external environmental conditions. Since these cultured plantlets are initially grown under controlled laboratory environments, they must undergo a transition phase before being transferred to greenhouses or open-field conditions. The goal of acclimatization is to increase the survival rate of tissue-cultured plantlets and ensure their successful growth under natural conditions. In a study conducted by Nazirah et al. (2021), well-rooted *E. grandis* × *E. urophylla* plantlets were acclimatized via coco peat and peat moss medium. Moreover, *E. microcorys* plantlets were acclimatized via a 1:1 combination of pine bark and vermiculite (Faria et al. 2021). In addition to the combination of media, some studies have combined three types of media for acclimatization. Shwe and Leung (2020) reported that a 1:1:1 blend of garden soil, sand, and compost was effective for adapting *E. bosistoana* plantlets. Similarly, *E. camaldulensis* plantlets successfully acclimatized when they were grown in a 2:1:1 mixture of peat moss, perlite, and vermiculite (Ahad et al. 2018). These studies show that a wide range of potting media can be utilized for acclimatizing *Eucalyptus* plantlets.

3.6 Callogenesis

In vitro callogenesis, which involves the formation of undifferentiated cell masses (calli), is a fundamental aspect of plant tissue culture. Several *Eucalyptus* species, including *E. cloeziana, E. microcorys, E. bosistoana,* and *E. nitens,* have been successfully subjected to callus culture via hypocotyl and cotyledonary leaves obtained from germinated seeds (Ayala et al. 2019; Shwe and Leung 2020; Faria et al. 2021; de Oliveira et al. 2022). Typically, callus induction is achieved via MS or WPM medium supplemented with plant growth regulators, most notably NAA and TDZ (de Oliveira et al. 2022). Once the callus is successfully induced, it is transferred to the regeneration medium to promote shoot development. The regeneration basal medium is often fortified with a combination of cytokinin and auxin, with cytokinin present at a higher concentration than auxin. For example, in a study conducted by Faria et al. (2021), callus regeneration of *E. microcorys* was achieved via WPM supplemented with 4.44 μM BAP and 0.27 μM NAA. Similarly, callus regeneration of *E. nitens* was conducted on MS medium supplemented with 2.5 μM BAP and 0.5 μM IAA (Ayala et al. 2019). Following successful regeneration, the callus-derived plantlets undergo a rooting stage before acclimatization and eventually transfer to the field.

4 Conclusion

Plant tissue culture has proven to be an effective alternative propagation method for *Eucalyptus* species, overcoming the limitations of conventional techniques. This approach ensures the production of genetically uniform, pest- and disease-free plantlets, making it a valuable tool for large-scale propagation. The success of *Eucalyptus* micropropagation depends on several key factors, including basal medium formulation, plant growth regulators, and culture conditions. Each stage of tissue culture, depending on the type of explant, requires a distinct basal medium formulation. Therefore, optimizing these formulations is paramount for establishing successful micropropagation protocols for *Eucalyptus*. However, further research is needed to address the existing challenges in micropropagation. Advances in genetic transformation hold a bright future for developing *Eucalyptus* varieties with improved traits such as disease resistance and high growth rates to increase the efficiency and sustainability of *Eucalyptus* production. As the field continues to evolve, a deeper understanding of the molecular and physiological mechanisms underlying tissue culture will pave the way for greater commercial applications. These advancements will ultimately contribute to the productivity and sustainability of *Eucalyptus* plantations, benefiting both the forestry and industrial sectors. In Southeast Asia, where land availability and timber demand are growing concerns, efficient micropropagation of *Eucalyptus* can support reforestation efforts and meet the region's increasing need for fast-growing, high-quality wood resources. Moreover, large-scale *Eucalyptus* cultivation plays a significant role in climate mitigation through carbon sequestration, helping to offset greenhouse gas emissions.

References

Afroze F, Douglas GC, Grogan H (2021) A comparative study on seed physiology and germination requirements for 15 species of *Eucalyptus*. Theor Exp Plant Physiol 33(4):411–425

Ahad A, John E, Maqbool A, Malik KA (2018) Development of efficient micropropagation system for *E. camaldulensis* with respect to age of explants. Pak J Agric Sci 55:23–27

Ahansal K, Aadel H, Udupa SM, Gaboun F, Abdelwahd R, Ibriz M, Iraqi D (2022) Effect of picloram and 2, 4-D on plant regeneration from mature and immature embryos of moroccan durum wheat varieties. J Plant Biotechnol 49(2):131–138

Almas I, Innocent E, Machumi F, Kisinza W (2021) Chemical composition of essential oils from *Eucalyptus globulus* and *Eucalyptus maculata* grown in Tanzania. Sci Afr 12:e00758

Animasaun DA, Adedibu PA, Afolabi SO, Abdulkareem KA, Ibrahim S, Krishnamurthy R (2024) Hazard assessment and cytogenotoxic effect of different concentrations of mercury chloride sterilant using the *Allium cepa* assay. Discov Toxicol 1(1):2

Avelar MLM, Souza DMSC, Macedo EH, Molinari LV, Brondani GE (2020) *In vitro* establishment of *Eucalyptus* and *Corymbia* species from epicormic shoots. Revista Árvore 44:e4427

Avelar M, Santana Costa Souza DM, Molinari LV, Fernandes SB, Tannure Faria JC, de Carvalho D, Leal Teixeira G, Brondani GE (2024) *In vitro* cloning of a selected *Eucalyptus pilularis* tree and genetic stability analysis of micropropagated plants. Bosque (Valdivia) 45(1):139–149

Ayala PG, Brugnoli EA, Luna CV, González AM, Pezzutti R, Sansberro PA (2019) *Eucalyptus nitens* plant regeneration from seedling explants through direct adventitious shoot bud formation. Trees 33:1667–1678

Bano AS, Khattak AM, Basit A, Alam M, Shah ST, Ahmad N, Gilani SAQ, Ullah I, Anwar S, Mohamed HI (2022) Callus induction, proliferation, enhanced secondary metabolites production and antioxidants activity of *Salvia moorcroftiana* L. as influenced by combinations of auxin, cytokinin and melatonin. Braz Arch Biol Technol 65:e22210200

Bidabadi SS, Jain SM (2020) Cellular, molecular, and physiological aspects of *in vitro* plant regeneration. Plants 9(6):702

Brunoni F, Rolli E, Polverini E, Spíchal L, Ricci A (2021) The adjuvant activity of two urea derivatives on cytokinins: an example of serendipitous dual effect. Plant Growth Regul 95:169–190

Cavallaro V, Pellegrino A, Muleo R, Forgione I (2022) Light and plant growth regulators on *in vitro* proliferation. Plants 11(7):844

Čmiková N, Galovičová L, Schwarzová M, Vukic MD, Vukovic NL, Kowalczewski PŁ, Bakay L, Kluz MI, Puchalski C, Kačániová M (2023) Chemical composition and biological activities of *Eucalyptus globulus* essential oil. Plants 12(5):1076

de Oliveira LS, Brondani GE, Molinari LV, Dias RZ, Teixeira GL, Gonçalves AN, de Almeida M (2022) Optimal cytokinin/auxin balance for indirect shoot organogenesis of *Eucalyptus cloeziana* and production of *ex vitro* rooted micro-cuttings. J For Res 33(5):1573–1584

Esquivel F, Castillo A, Bentancor M, Ceppa M, Rogel L, Bonilla MB, Balmelli G, Dalla-Rizza M (2024) Potential of metatopoline in the *in vitro* multiplication and rooting of *Eucalyptus globulus* Labill. Clones CERNE 30:e-103413

Faria JCT, Terra JAP, Molinari LV, Delarmelina WM, Ribeiro-Kumara C, de Sena Neto AR, de Carvalho D, Brondani GE (2021) Use of polylactic acid microvessel to obtain microplantlets of *Eucalyptus microcorys* through indirect organogenesis. 3 Biotech 11(8):364

Fazili MA, Bashir I, Ahmad M, Yaqoob U, Geelani SN (2022) *In vitro* strategies for the enhancement of secondary metabolite production in plants: A review. Bull Natl Res Cent 46(1):35

Frade SR, Souza DMSC, Fernandes SB, Avelar MLM, Molinari LV, Gonçalves DS, Magalhães TA, Brondani GE (2023) Spectral quality influence on *in vitro* morphophysiological responses of *Eucalyptus dunnii* Maiden and *Eucalyptus grandis* W. Hill ex Maiden × *E. urophylla* ST Blake. N Z J For Sci 53:3

Gantait S, Mahanta M (2021) Picloram-induced enhanced callus-mediated regeneration, acclimatization, and genetic clonality assessment of gerbera. J Genet Eng Biotechnol 19(1):175

Gomes GLB, Scortecci KC (2021) Auxin and its role in plant development: structure, signalling, regulation and response mechanisms. Plant Biol 23(6):894–904

Hameg R, Arteta TA, Landin M, Gallego PP, Barreal ME (2020) Modeling and optimizing culture medium mineral composition for *in vitro* propagation of *Actinidia arguta*. Front Plant Sci 11:554905

Heck MA, Lüth VM, van Gessel N, Krebs M, Kohl M, Prager A, Joosten H, Decker EL, Reski R (2021) Axenic *in vitro* cultivation of 19 peat moss (*Sphagnum* L.) species as a resource for basic biology, biotechnology, and paludiculture. New Phytol 229(2):861–876

Isah T, Umar S (2020) Influencing *in vitro* clonal propagation of *Chonemorpha fragrans* (moon) Alston by culture media strength, plant growth regulators, carbon source and photo periodic incubation. J For Res 31:27–43

Karakas FP (2020) Efficient plant regeneration and callus induction from nodal and hypocotyl explants of goji berry (*Lycium barbarum* L.) and comparison of phenolic profiles in calli formed under different combinations of plant growth regulators. Plant Physiol Biochem 146:384–391

Karet R, Nakhooda M, Jones NB, Hills PN (2021) Optimisation of micropropagation protocols for temperate *Eucalypt* hybrids in South Africa, with a focus on auxin transport proteins. South For J For Sci 83(4):254–263

Khedhri S, Polito F, Caputo L, Khammassi M, Dhaouadi F, Amri I, Hamrouni L, Mabrouk Y, Fratianni F, Nazzaro F, De Feo V (2024) Antimicrobial, herbicidal and pesticidal potential of Tunisian *Eucalyptus* species: Chemoprofiling and biological evaluation. Heliyon 10(9):e29905

Kumar M, Sarvade S, Kumar R, Kumar A (2024) Pre-sowing treatments on seeds of forest tree species to overcome the germination problems. Asian J Environ Ecol 23(5):1–18

Mehbub H, Akter A, Akter MA, Mandal MSH, Hoque MA, Tuleja M, Mehraj H (2022) Tissue culture in ornamentals: cultivation factors, propagation techniques, and its application. Plants 11(23):3208

Molinari LV, Souza DMSC, Avelar MLM, Fernandes SB, Gonçalves DS, Faria JCT, de Carvalho D, Brondani GE (2021) Effects of chemical sterilization of the culture media, porous membranes and luminosity on *in vitro* culture of *Eucalyptus grandis* × *Eucalyptus urophylla*. J For Res 32(4):1587–1598

Nazirah A, Nor-Hasnida H, Mohd-Saifuldullah AW, Muhammad-Fuad Y, Ahmad-Zuhaidi Y, Rozidah K (2021) Development of an efficient micropropagation protocol for *Eucalyptus* hybrid (*E. urophylla* × *E. Grandis*) through axillary shoot proliferation. J Trop For Sci 33(4):391–397

Qin C, Zeng H, Liu B, Zhu J, Wang F, Wang S, Liang C, Huang C, Ma J, Yao (2022) Efficient removal of residual lignin from *Eucalyptus* pulp via high-concentration chlorine dioxide treatment and its effect on the properties of residual solids. Bioresour Technol 360:127621

Sharma V, Ankita, Karnwal A, Sharma S, Kamal B, Jadon VS, Gupta S, Sivanasen I (2023) A comprehensive review uncovering the challenges and advancements in the *in vitro* propagation of *Eucalyptus* plantations. Plants 12(17):3018

Shwe SS, Leung DW (2020) Plant regeneration from *Eucalyptus bosistoana* callus culture. In Vitro Cell Dev Biol Plant 56:718–725

Singh R, Dutta RK, Naik DV, Ray A, Kanaujia PK (2021) High surface area *Eucalyptus* wood biochar for the removal of phenol from petroleum refinery wastewater. Environ Chall 5:100353

Sohrabi S, Abdollahi MR, Mirzaie-Asl A, Koulaei HE, Aghaeezadeh M, Seguí-Simarro JM (2021) A refined method for ovule culture in sugar beet (Beta vulgaris L.). Plant Cell, Tissue and Organ Culture 146(2):259–267

Souza DMSC, Fernandes SB, Avelar MLM, Frade SRDP, Molinari LV, Gonçalves DS, Pinto JEBP, Brondani GE (2020) Light quality in micropropagation of *Eucalyptus grandis* × *Eucalyptus urophylla*. Scientia Forestalis 28(127):e3329

Terceros GC, Resentini F, Cucinotta M, Manrique S, Colombo L, Mendes MA (2020) The importance of cytokinins during reproductive development in Arabidopsis and beyond. Int J Mol Sci 21(21):8161

Tien LH, Chac LD, Oanh LTL, Ly PT, Sau HT, Hung N, Thanh VQ, Doudkin RV, Thinh BB (2020) Effect of auxins (IAA, IBA and NAA) on clonal propagation of *Solanum procumbens* stem cuttings. Plant Cell Biotechnol Mol Biol 21(55–56):113–120

Trueman SJ, Hung CD, Wendling I (2018) Tissue culture of *Corymbia* and *Eucalyptus*. Forests 9(2):84

Vahdati K, Sadeghi-Majd R, Sestras AF, Licea-Moreno RJ, Peixe A, Sestras RE (2022) Clonal propagation of walnuts (*Juglans* spp.): a review on evolution from traditional techniques to application of biotechnology. Plants 11(22):3040

Vylíčilová H, Bryksová M, Matušková V, Doležal K, Plíhalová L, Strnad M (2020) Naturally occurring and artificial N9-cytokinin conjugates: from synthesis to biological activity and back. Biomolecules 10(6):832

Wang Y, Wu Z, Li X, Shang X (2024) Regeneration and genetic transformation in *Eucalyptus* species, current research and future perspectives. Plants 13(20):2843

Yang W, Cortijo S, Korsbo N, Roszak P, Schiessl K, Gurzadyan A, Wightman R, Jönsson H, Meyerowitz E (2021) Molecular mechanism of cytokinin-activated cell division in Arabidopsis. Science 371(6536):1350–1355

Zhu W, Zhang Y, Wang X, Wu Y, Han M, You J, Jia C, Kim J (2022) Aerogel nanoarchitectonics based on cellulose nanocrystals and nanofibers from eucalyptus pulp: preparation and comparative study. Cellulose 29(2):817–833

Wildlife in Planted Eucalypt Forests: Diversity and Ecological Response to Habitat Alteration

Chong Leong Puan, Amera Natasha Mah Muhammad Adam Mah, and Al-Kautsar Hidayanto Abdul Rahim

Abstract Large-scale planting of native and nonnative trees (e.g., eucalypts) in Malaysia began approximately 15 years ago. These forest plantations are highly economically important with respect to supplying timber locally and internationally. This national initiative also reduces logging pressure on natural forests. However, owing to differences in habitat structure and floristic composition, planted forests may be able to support only some but not all native wildlife species, as do natural forests. This chapter reviews 56 published works on eucalypt plantations in different countries to identify which animal species, both vertebrates and invertebrates, have benefited from these plantations or not. This finding also explains why certain species thrive in plantations while others do not, from an ecological point of view. This review is essential considering the increasing concerns over the role of forest plantations, specifically eucalypt planting, in biodiversity conservation. Recommendations are given on planting systems and practices that increase native biodiversity, which is essential if the ultimate aim of establishing planted forests is to improve their role in meeting the goals of sustainable development and biodiversity conservation in the long run.

1 Introduction

Malaysia is known globally for its rich tropical forest biome, within which unique and rich species assemblages are found. With such high biodiversity, Malaysia has long been recognized as one of the top 12 countries with mega-biodiversity on the

C. L. Puan (✉)
Faculty of Forestry and Environment, Universiti Putra Malaysia, Serdang, Selangor, Malaysia

Institute of Tropical Forestry and Forest Products, Universiti Putra Malaysia, Serdang, Selangor, Malaysia
e-mail: chongleong@upm.edu.my

A. N. M. Muhammad Adam Mah · A.-K. H. Abdul Rahim
Faculty of Forestry and Environment, Universiti Putra Malaysia, Serdang, Selangor, Malaysia

S. Mohd Razali et al. (eds.), *Sustainable Eucalyptus Forestry in Southeast Asia*, Forestry Sciences 88, https://doi.org/10.1007/978-3-032-21343-3_2

basis of the National Biodiversity Index (Shevade et al. 2017) and is a biodiversity hotspot (Zeng et al. 2018; Feng et al. 2021) and global ecoregion (Loucks et al. 2008). Within complex tropical forest ecosystems, organisms occupy distinct ecological niches, such as different strata of rainforests or varying vegetation zones in riparian habitats. All these factors contribute to the complex ecological interactions among organisms and between biotic communities and abiotic elements, creating cycles that facilitate the transfer of nutrients and energy, which ultimately sustain essential ecological functions and services for humanity.

The abundance of natural resources in tropical regions inevitably leads to increased exploitation. In Southeast Asia, the relative deforestation rate is the highest of any major tropical region. It is estimated that three-quarters of its original forests and up to 42% of its biodiversity will be lost by 2100 (Sodhi et al. 2004). For many decades, people have been harvesting timber and nontimber forest products, primarily for commercial purposes. The impact of forest exploitation varies depending on management practices, ranging from the removal of large trees, the destruction of complex forest structures and microhabitats, and soil erosion. The degree of habitat disturbance caused by such forest conversion alters tropical biotic communities in terms of both species composition (Punia et al. 2022) and population size (Mitchell et al. 2022), and it often results in disruptions in biogeochemical cycles (Sullivan et al. 2019) and climate (Wood et al. 2012).

In fact, considering Peninsular Malaysia alone, forested areas decreased annually by 0.38% from 2007 to 2016 (Omar and Misman 2018). In view of the land use pressure on natural forests, planted forests serve as an alternative option to supply timber to downstream industries. In Malaysia, the establishment of forest plantations dates back to the 1930s in West Malaysia and the 1970s in East Malaysia. Over the last 15 years, many large-scale commercial forest plantations have been gradually established across the country, featuring both native and nonnative species ranging from eucalypts (*Eucalyptus* spp.), mangium (*Acacia mangium*), Kelampayan (*Neolamarckia cadamba*), Teak (*Tectona grandis*), and Sentang (*Azadirachta excelsa*) to Binuang (*Octomeles sumatrana*).

2 Wildlife Composition in Planted Forests

As global timber extraction increasingly shifts to planted rather than natural forests, the response of animals to planted forests, particularly for species that rely strictly on natural forest habitats for their survival, remains unexplored. Unlike natural tropical forests, planted forests are often monocultures with reduced floristic species compositions and less complex habitat structures. As a result, some wildlife species may thrive in planted forests, whereas others may not (Cannell 1999; Barlow et al. 2007a; Intachat et al. 1999; Jamil et al. 2020).

Past studies have shown that the wildlife composition of planted plantations is influenced by both local and landscape factors (Duff et al. 1984; Stuebing and Gasis 1989; Styring et al. 2011). In Sabah, bird diversity increased with the age of the mangium plantations; however, these results were only true for common and smaller species, not large, specialized ones (Styring et al. 2011). Similarly, the abundances of small mammals (such as Whitehead's Spiny Rat (*Maxomys whiteheadi*), Common Treeshrew (*Tupaia glis*), and Large Treeshrew (*Tupaia tana*)) and larger terrestrial mammals (e.g., deer, civets, Bearded Pig (*Sus barbatus*), and Leopard Cat (*Felix bengalensis*)) are greater in nonnative tree plantations in Sabah, particularly those located close to natural forests (Duff et al. 1984). In the case of invertebrates, certain insects (e.g., wood borers, leaf-eating beetles, gall wasps, termites, and plant lice) have successfully adapted to planted plantations and even caused severe damage (Schabel et al. 1999). For example, Kelampayan plantations are often infested with wood borers, causing deterioration in wood quality and timber value (Nair 2001). In Sarawak, approximately 70% of wood borer attacks occur in major Kelampayan plantations (Chai et al. 2010). Another study that compared geometrid moth diversity in four plantations planted with indigenous and nonnative species in Sabah and Peninsular Malaysia reported that the diversity was greater in areas planted with indigenous species (Intachat et al. 1999).

2.1 Research on the Wildlife of Eucalypt Plantations

Despite originating from Australasia, eucalypts have now been widely planted on other continents because of their fast-growing and adaptive nature as well as the high-value timber or raw wood materials they produce (Zaiton et al. 2018). Given the increasing attention given to commercial eucalypt planting in Malaysia and elsewhere, as highlighted in other chapters of this book, this chapter reviews published studies from the past three decades on the responses of wildlife to eucalypt plantations. We will first present the main results based on different groups of animals and countries (Table 1), and then summarize how various factors affect the types of wildlife found in eucalypt plantations.

A total of 56 published works revealed that the majority of the research was conducted in South America (15 studies), followed by Europe (15), Asia (11), Australia (10), Africa (4), and North America (1) (Table 1, with details presented in the appendix). Only eight of these studies were conducted in Malaysia. In terms of focal animal groups, these studies covered either various taxa (7.1%) or certain invertebrates (37.5%) and vertebrates (birds, 33.9%; large and medium-sized mammals, 8.9%; small mammals, 8.9%; amphibians, 1.8%; and fish, 1.8%). Among the species planted in the study areas were *Eucalyptus globulus, E. grandis, E. deglupta, E. urograndis, E. urophilia, E. nitens, E. pilularis, E. tereticornis, E. pellita,* and *E. saligna*.

Table 1 Summary of studies on the response of different taxa to eucalypt plantations

Study location	Taxa	Effect	Responses	Source
Brazil	Macroinvertebrates	–	Many invertebrate pest species are harmful to plantations	Anjos and Majer (2003)
		–	Primary forests had higher richness and abundance than plantations	Hawes et al. (2009)
		+/–	Some taxa densities similar or higher in plantations (+); social insects lower (–)	Pellens and Garayx (1999)
		+/–	Higher species richness in older plantations (+); more generalists (–)	Suguituru et al. (2011)
		+/–	Higher beetle diversity in eucalypt than pine plantations (+); some aggressive pests (–)	Flechtmann et al. (2001)
		+/–	Lower species number in plantations (–); older plantations better habitat (+)	Gries et al. (2012)
	Birds	–	Lower richness and abundance in plantations vs. forest fragments	Marsden et al. (2001)
		–	Lower specialist bird species in plantations	Barlow et al. (2007b)
		+	Similar bird richness in old eucalypt plantations and native forest	Lopes et al. (2015)
	Mammals	–	Fewer species and individuals in plantations than in a natural savanna	Martin et al. (2012b)
		+	Some species are abundant; composition varies by plantation age	Timo et al. (2015)
		+	Similar species richness and abundance between plantations and forest fragments	Gheler-Costa et al. (2012)
		+	Some mammal diversity is supported in plantations with nearby natural habitat	Piña et al. (2019)
	Insects, birds, amphibians, bats	–	Lower bird, butterfly, and amphibian richness in plantations	Barlow et al. (2007a)
Chile	Birds	+	Similar richness and abundance between the native forest and the abandoned plantations	Fontúrbel et al. (2016)
Mexico	Macroinvertebrates	+	Higher spider richness in eucalypt plantations	Corcuera et al. (2010)

(continued)

Table 1 (continued)

Study location	Taxa	Effect	Responses	Source
Puerto Rico	Amphibians	+	No density difference between plantations and native forest	Fogarty and Vilella (2003)
Portugal	Macroinvertebrates	–	Poor food quality of eucalypt leaves, negative effects on streams	Canhoto and Graça (1995)
		–	Monoculture eucalypts reduce stream invertebrate diversity	Abelho and Graca (1996)
		+/–	Lower arthropod abundance in plantations (–); leaf litter quality affects fauna (+/–)	Zahn et al. (2009)
	Birds	–	Lower forest bird richness in plantations vs. oak and pine forests	Proença et al. (2010)
	Mammals	–	Negative correlation of red fox occupancy with plantations	Cruz et al. (2015)
		+	Shrub cover affects abundance; seasonal variation	Carrilho et al. (2017)
Portugal-Spain (Iberian Peninsula)	Macroinvertebrates	–	Mixed effects on stream invertebrates; short rotations reduce habitat quality	Graça et al. (2002)
Spain	Macroinvertebrates	–	Lower shredder abundance in plantations	Larrañaga et al. (2006)
		–	Mixed community impacts	Larrañaga et al. (2009)
		–	Diversity decreases as eucalypt cover increases; a shift in trophic structure	Cordero-Rivera et al. (2017)
		+	No drastic impacts on stream community structure	Pozo et al. (1997)
		+	Higher invertebrate abundance in alder vs. eucalypt litter	Basaguren and Pozo (1994)
	Birds	–	Lowest richness and abundance in eucalypt plantations	Calviño-Cancela (2013)
		+	Higher richness and abundance in native forests; some species are exclusive to plantations	Goded et al. (2019)
China	Fish	–	Mortality caused by hypoxia and black water toxicity from eucalypt plantations	Yang et al. (2019)
	Birds	–	Lower species richness and capture rate in plantations vs natural forests	Zou et al. (2014)

(continued)

Table 1 (continued)

Study location	Taxa	Effect	Responses	Source
Malaysia	Macroinvertebrates	+	High species diversity in plantations is comparable to that of natural forests	Chey et al. (1997)
	Birds	–	Lower plant and bird diversities in plantations	Zaiton et al. (2020)
		+	Similar richness in plantations and pristine forests	Soh et al. (2006)
		+	Presence of bird species, including the little bronze-cuckoo, in plantations	Sheldon et al. (2009)
	Mammals	+	Plantations support small mammals, though canopy species remain low	Stuebing and Gasis (1989)
		–	Lower abundance in plantations, but some species are present at the edges	Duff et al. (1984)
	Insects, mammals	–	Presence of pests damaging plantations	Chung et al. (2015)
	Various wildlife	+	Wildlife corridors are recommended to support biodiversity	Hon and Shibata (2013)
Sri Lanka	Birds	–	Lower species diversity in eucalypt plantations compared to other habitats	Kottawa-Arachchi and Gamage (2015)
Australia	Macroinvertebrates	–	Lower native snail and millipede diversity; more introduced species in plantations	Bonham et al. (2002)
	Insects	–	Lower richness than forest remnants; monocultures favour herbivores	Cunningham et al. (2008)
	Birds	+	Abundance higher than farmland but lower than native forests; varied guild presence	Loyn et al. (2007)
		+	Mature eucalypts support hollow-using species	Koch et al. (2009)
		+	47% increase in bird species over 11 years; supports insectivores and nectarivores	Law et al. (2014)
		+	Improved bird density vs cleared paddocks; supports declining woodland species	Kavanagh et al. (2007)
		+	Riparian plantations support more species and feeding guilds than dryland plantations	Hsu et al. (2010)

(continued)

Table 1 (continued)

Study location	Taxa	Effect	Responses	Source
	Bats, birds	+	Provides habitat for at-risk bird species; bat and bird abundance are higher near remnant vegetation	Lindenmayer and Hobbs (2004)
	Koalas	+	Young plantations near remnant vegetation aid koala habitat restoration	Kavanagh and Stanton (2012)
	Amphibians, reptiles, birds, mammals	–	Animals use plantations less than remnant vegetation, more than open pasture	Hobbs et al. (2003)
Africa	Macroinvertebrates	–	Pest damage to *E. nitens* plantations	Wingfield et al. (2008)
	Birds	–	Lower species richness than natural forests, though some benefit from well-managed plantations	John and Kabigumila (2011)
Madagascar	Birds	+	Higher forest bird richness than other land uses; an important habitat mosaic	Martin et al. (2012a)
	Bats	+	Bats pollinate eucalypt flowers	Picot et al. (2007)

Note: + positive response; – negative response

3 Response of Different Taxa to Eucalyptus Planting

In their native habitats in Australia, eucalypt plantations facilitated the reinvasion of some invertebrates into farmlands from adjacent bush remnants. A comparison of ground-dwelling invertebrates between eucalypt plantations and native forests in Tasmania revealed that the abundance of introduced land snails and velvet worms was greater in the former than in the latter, whereas native land snail and millipede diversity was greater in the native forests (Bonham et al. 2002). In eucalypt plantations in southwestern Australia, there are more leaf-feeding moths and beetles but fewer ants, bees, evanioid wasps, and predatory beetles than in adjacent forest remnants (Cunningham et al. 2008).

Like invertebrates, the responses of vertebrates in Australia also vary according to their habitat requirements. In Tasmania, the presence and abundance of hollows in mature eucalypts are important for maintaining biodiversity, particularly for hollow-nesting birds (Koch et al. 2009), although nesting may not always occur (Lindenmayer and Hobbs 2004). In eastern Australia, eucalypt plantations were found to be useful habitats for at least 10 declining woodland-dependent bird species. Lindenmayer and Hobbs (2004) also concluded that eucalypt plantations provide habitat or resources for a number of bird species that are at risk of being threatened. In southeastern Australia, the abundance of forest and woodland birds, as well as birds of certain feeding guilds, was greater in eucalypt plantations than in cleared farmlands, although it was lower than that in native forests (Loyn et al. 2007). In a related study, bird communities in mixed eucalypt–shrub plantations in southeastern Australia closely resembled those in native forest remnants and woodlands (Kavanagh et al. 2007). Moreover, bird and bat abundances were greater in plantations adjoining remnant vegetation than in those with isolated vegetation. In fact, a decade of eucalypt plantation growth in northern New South Wales resulted in a 47% increase in bird species (Law et al. 2014). In the case of mammals, although tree hollows in eucalypt plantations are not used by arboreal marsupials (Lindenmayer and Hobbs 2004), plantations located in close proximity to forest remnants have the potential to restore habitat for koalas (Kavanagh and Stanton 2012).

In reforested areas of Brazil, where eucalypts constitute one of the two major exotic trees planted, the diversity of certain arthropods, e.g., dung beetles (Gries et al. 2012), scolytid beetles (Flechtmann et al. 2001), and generalist ants (Suguituru et al. 2011), was found to be greater than that of other habitat types. Other invertebrates, e.g., some soil arthropods (Pellens and Garayx 1999), which are associated with the quantity and quality of eucalypt leaf litter, similar to the case of stream arthropods in Spain (Larrañaga et al. 2006; Larrañaga et al. 2009), exhibited the opposite response to planted forests. A similar negative trend was also observed for terrestrial small mammals in southern and southeastern Brazil, with most generalists being recorded (Gheler-Costa et al. 2012; Martin et al. 2012b). Among larger vertebrates in the country, generalist mammals thrive in eucalypt plantations, whereas the occurrence of certain mammal species depends on plantation age (Timo et al. 2015). Similarly, birds represented more generalists than specialists in eucalypt plantations. More specifically, feeding guilds presented a high abundance of nectarivores, a low number of large arboreal frugivores and granivores, and absence of arboreal granivores in northeastern Brazil (Barlow et al. 2007b).

In Portugal, widespread eucalypt monocultures have been found to affect stream invertebrate communities (Canhoto and Graça 1995; Abelho and Graca 1996), and maintaining riparian corridors of native vegetation is important for reducing such impacts (Abelho and Graca 1996). In another study on ground macroarthropods in Portugal, the lower soil moisture in eucalypt plantations led to a lower abundance of arthropods than in riparian habitats, which can be linked to leachates from eucalypt leaves (Zahn et al. 2009). In terms of birds, Proença et al. (2010) recorded 28 bird species in eucalypt stands, including 11 forest species, in northwestern Portugal, although the number of forest species remained lower than that in oak and pine forests. In mammals, the abundance of small mammals was associated with the availability of shrub cover in eucalypt plantations in central-western Portugal (Carrilho et al. 2017). In a study of three carnivores in central-eastern Portugal, the occupancy of Red Fox (*Vulpes vulpes*) was found to be influenced by eucalypt cover (Cruz et al. 2015).

Many studies have shown that the diversity of fauna in eucalypt plantations is generally not comparable to that in natural forests (Poore and Fries 1985; Cannell 1999; Marsden et al. 2001; Bonham et al. 2002; Loyn et al. 2007; Gries et al. 2012; Calviño-Cancela 2013; Zou et al. 2014; Cordero-Rivera et al. 2017). For example, a study in Sabah, Malaysia, that compared small mammal diversity across several plantation types (including eucalypts) and lowland forests revealed that canopy species were absent in the plantations, despite the presence of some forest species (Stuebing and Gasis 1989). Similarly, studies conducted in Brazil have also demonstrated that mammal (Zanzini et al. 2017) and dung beetle (Gries et al. 2012) species richness is greater in natural forests than in eucalypt plantations, although certain arboreal species, specifically mammals, thrive in plantations (Zanzini et al. 2017). Similar results indicating greater bird richness were reported in forests in Galicia, Spain (Calviño-Cancela 2013; Goded et al. 2019) and the Usambara Mountains in Tanzania (John and Kabigumila 2011) than in eucalypt plantations, although some forest species were recorded exclusively in eucalypt plantations (John and Kabigumila 2011; Goded et al. 2019).

Several studies on the quality and quantity of leaf litter in northern Spain have indicated that stream macroinvertebrates, particularly shredders (*Nemoura* spp. and *Echinogammarus* spp.), are less abundant in eucalypt plantations than in native forests (Graça et al. 2002; Larrañaga et al. 2006, 2009; Cordero-Rivera et al. 2017). A number of factors contribute to these patterns, including lower habitat complexity (Pellens and Garayx 1999; Fogarty and Vilella 2003; Suguituru et al. 2011; Timo et al. 2015; Piña et al. 2019; Zaiton et al. 2020), limited food sources, e.g., insect prey (Calviño-Cancela 2013; Cordero-Rivera et al. 2017; Goded et al. 2019), seasonal changes in weather (Pellens and Garayx 1999), silvicultural practices (Gheler-Costa et al. 2012; Timo et al. 2015), and soil hydrophobicity (Graça et al. 2002; Zahn et al. 2009) in eucalypt plantations.

On the other hand, comparative analyses of planted eucalypts and other habitat types have revealed negligible differences (Hobbs et al. 2003), increases (Chey et al. 1997; Loyn et al. 2007; Corcuera et al. 2010; Martin et al. 2012a), and decreases in faunal diversity (Proença et al. 2010; Kottawa-Arachchi and Gamage 2015; Cordero-Rivera et al. 2017). This again implies that some fauna species may be able to adapt to eucalypt monocultures, whereas others cannot. In Madagascar, the endemic Madagascan Fruit Bat (*Eidolon dupreanum*) has been observed feeding on the flowers of eucalypts (Picot et al. 2007). Some native species may thrive, e.g., moths in Sabah, Malaysia (Chey et al. 1997), whereas others become pests in eucalypt plantations in some countries (e.g., Flechtmann et al. 2001; Nair 2001; Anjos and Majer 2003; Wingfield et al. 2008; Chung et al. 2015). In some studies, greater faunal diversity in eucalypt plantations is associated with edge habitats (Duff et al. 1984; Hobbs et al. 2003; John and Kabigumila 2011) or the age of planting (Law et al. 2014; Timo et al. 2015; Zaiton et al. 2020), which is also linked to the abundance of undergrowth (Marsden et al. 2001; Suguituru et al. 2011; Timo et al. 2015; Carrilho et al. 2017). In fact, Lopes et al. (2015) reported that, despite being exotic, old eucalypt stands of more than 70 years in Brazil can complement native species plantations in maintaining bird diversity. Other studies also concurred with the idea of having eucalypt planting as a complementary approach to maintain the richness of mammals (Piña et al. 2019) and other taxa (Barlow et al. 2007a) in Brazil.

4 The Way Forward for Biodiversity Conservation

A review of 56 journal articles indicates that species richness, abundance, and/or density are often greater in native forests than in eucalyptus plantations, particularly for birds, bats, some invertebrates, reptiles, and some amphibians. The responses of specialist and generalist species differ between these habitat types. Most of the studies were performed in countries where *Eucalyptus* spp. are widely planted, i.e., Brazil, Australia, Spain, and Portugal. Comparing eucalypt plantations to other human-made habitats showed different results based on factors like habitat layout, food supply, living space, age of the plantation, nearby habitats, and seasonal weather changes. Some bird and mammal species occurring in native habitats are absent in eucalypt plantations, and vice versa.

Well-managed eucalypt plantations, including those with fast-growing and high-value species, may benefit wildlife (Cannell 1999; Rahim et al. 2023), including birds (John and Kabigumila 2011) and mammals (Timo et al. 2015). Several options are available to minimize the impacts of eucalypt planting on native fauna while enriching their habitats, i.e., complementary, alternative, or suboptimal habitats. These include maintaining habitat structural complexity (Chey et al. 1997; Soh et al. 2006), preserving riparian corridors (Abelho and Graca 1996) or buffer zones to ensure habitat connectivity (Soh et al. 2006; Kavanagh et al. 2007; Graça et al. 2002; Hon and Shibata 2013), carefully selecting planting locations to create the habitat matrices needed by certain species (Lindenmayer and Hobbs 2004; Cunningham et al. 2008; Gries et al. 2012; Martin et al. 2012a, b; Law et al. 2014), and planting mixed species, especially indigenous trees (Kavanagh et al. 2007). In addition, planting eucalypts on cleared farmland allows some fauna, e.g., invertebrates, to reinvade from adjacent bush remnants, which helps reduce the risk of local extinctions, as in the case of Australia (Bonham et al. 2002). Woodland-dependent bird species can also benefit from the replanting of eucalyptus trees in cleared farmlands. Mature eucalypt plantations can provide hollows for hollow-using or cavity-nesting species, although this benefit is yet to be explored elsewhere, except in Australia. Additionally, older plantations should be repurposed to establish new wildlife corridors (Hon and Shibata 2013) through replanting or restoring damaged buffers, especially along riverbanks. Abandoned eucalyptus plantations may also provide suitable habitat for some native species if native understorey vegetation is present, as demonstrated in many studies.

Integrated pest management (Anjos and Majer 2003) should be an effective approach to address the emergence and infestation of pests in plantations. Other potential threats to fauna, such as the cutover of eucalypt plantations in China, which has caused water pollution and killed certain fish species due to acute hypoxia and toxicity, can be mitigated by creating riparian buffer strips to reduce such human impacts on streams and rivers (Yang et al. 2019). The association between eucalypt planting and soil hydrophobicity (Graça et al. 2002; Zahn et al. 2009) has impacted some invertebrates in Europe, but its effects on other taxa have not been carefully assessed in regions such as Southeast Asia. However, certain natural areas with native plants, like riparian corridors, play a key role in lessening the negative effects of eucalypt plantations on insect communities, as shown in Europe.

In view of the expected expansion of eucalypt plantations in tropical developing countries, attempts should be made to meet the growing demand for timber and raw biomaterials without causing adverse impacts on native biodiversity. Ideally, strategies should aim to enrich biodiversity within these monocultures, as discussed earlier. This goal is particularly crucial considering the rich biodiversity found in tropical countries such as Malaysia, as discussed at the beginning of this chapter. While many past studies have indicated that wildlife diversity in eucalypt plantations is generally lower than that in native forests, some forest species persist or are found exclusively in some plantations (John and Kabigumila 2011; Goded et al. 2019; Piña et al. 2019). In Malaysia, such research is still in its infancy. A recent study by Rahim et al. (2023) documented several elusive species, including the Asian Golden Cat (*Catopuma temminckii*), Sunda Pangolin (*Manis javanica*), Large Indian Civet (*Viverra zibetha*), Sun Bear (*Helarctos malayanus*), and Serow (*Capricornis sumatraensis*), within eucalypt plantations. However, whether these native forest species can persist in such plantations and how they use this man-made habitat warrants further investigation. Hence, further studies are needed to examine how forest species respond to these plantations, including species that may cause damage to eucalypts and other commercially planted species. Such research will support the development of better-managed planted forests by improving our understanding of wildlife responses, their potential ecological roles in these man-made habitats (e.g., Duff et al. 1984), and the reciprocal influence of these habitats on wildlife (Sheldon and Styring 2011).

Competing Interests The authors declare that they have no conflicts of interest relevant to the content of this chapter.

Appendix: Global Studies on Animal Diversity and Species Composition in Eucalypt Plantations

Focus continent/ country	Species planted	Focal taxa	Major findings	Source
Brazil	*Eucalyptus* sp.	Insects	More than 200 invertebrate species were harmful to plantations. More than 40 leaf-eating beetle species are reported as pests, including chrysomelids, weevils, spring beetles, and jewel beetles. The most important pest species were *Costalimaita ferruginea*, *Gonipterus* spp., *Lampetis* spp., and *Bolax flavolineata*.	Anjos and Majer (2003)
Brazil (North)	*E. urograndis*	Moths	Primary forests had significantly higher species richness and abundance compared to secondary forests and eucalypt plantations. Species community structure and composition variation did not correlate with forest structure variables, although canopy cover, basal area of trees, and lianas were important factors identified. Differences in moth communities were determined by the distribution of host-plant resources instead of physical vegetation structure.	Hawes et al. (2009)
Brazil (Northeast)	*E. urograndis*	Various fauna	Eucalypt plantations had a lower species richness of birds, fruit-feeding butterflies, and leaf-littering amphibians compared to native forests. Highly vagile species (orchid bees and bats) showed idiosyncratic responses to both habitat types.	Barlow et al. (2007a)
Brazil (East)	*E. grandis*	Edaphic macroarthropods	Seven-year-old eucalypt plantations had the same and/or higher densities of some taxa compared to primary forests, although significantly higher abundances of some social insects and nonsocial macroarthropods were found in the forests due to the quality and quantity of leaf litter. During the rainy season (February), eucalypt plantations observed a reduced abundance of social insect species. Reduced abundance of nonsocial macroarthropods was observed in November and February in the plantations compared to forests in February and June.	Pellens and Garayx (1999)

Focus continent/ country	Species planted	Focal taxa	Major findings	Source
Brazil (South)	*E. grandis*, *E. urophilia*, *E. urograndis*	Small mammals	Fewer types and numbers of small mammals were found in eucalypt plantations than in natural secondary-growth Cerrado (savanna woodland) areas. Mostly generalist species (four species from Didelphimorphia and 10 species from Rodentia) were recorded. Of all captures, 81.8% were represented by Yellow Pygmy Rice Rat (*Oligoryzomys flavescens*), Black-footed Pygmy Rice Rat (*Oligoryzomys nigripes*), and Delicate Vesper Mouse (*Calomy tener*).	Martin et al. (2012b)
Brazil (South)	*E. grandis*, *E. saligna*	Leaf-litter ants	Species richness varied across *Eucalyptus* plantations with different understorey structures. Higher species richness was found in older, developed, and subdeveloped understorey (79 and 76 species, respectively) compared to young, managed understorey (61 species). Plantations are likely to support more generalist species and fewer specialist species.	Suguituru et al. (2011)
Brazil (South)	*E. grandis*	Ambrosia and bark beetles	Beetle diversity was higher in eucalypt than in pine plantations. The number of beetles found in the warm season was double that of the cold season. Approximately 75 beetle species were found in eucalypt stands, with *Hypothenemus eruditus*, *Xylosandrus retusus*, *Hypothenemus obscurus*, *Xyleborus ferrugineus*, and *Microcorthylus minimus* being the most abundant. *Premnobius cavipennis* and *Xyleborus ferrugineus* were the most aggressive scolytids in eucalypt stands, as they attacked healthy standing trees and saplings.	Flechtmann et al. (2001)

Focus continent/ country	Species planted	Focal taxa	Major findings	Source
Brazil (Southeast)	*Eucalyptus* sp.	Medium and large mammals	The most abundant species found in eucalypt plantations were omnivorous Manned-wolf (*Chrysocyon brachyurus*) and Crab-eating Fox (*Cerdocyon thous*). Species exclusively found in certain stand ages (i.e., Northern Tiger Cat (*Leopardus tigrinus*) and Collared Peccary (*Pecari tajacu*) in 0–1-year-old; Tapeti (*Sylvilagus brasiliensis*) in 2–3-year-old; Yellow Armadillo (*Euphractus sexcinctus*), South American Coati (*Nasua nasua*), Tayra (*Eira Barbara*), and Puma (*Puma concolor*) in 4–5-year-old planting). Armadillo (*Dasypus* spp.) and Giant Ant-eater (*Myrmecophaga tridactyla*) were also abundant in all stand ages. European Hare (*Lepus* spp.) was abundant in early plantation stage (0–1-year-old). Omnivores and insectivores were abundant in 2–5-year-old plantations.	Timo et al. (2015)
Brazil (Southeast)	*E. grandis*, *E. urophila*, *E. urograndis*	Small mammals	No difference in terms of species richness and relative abundance between semideciduous remnant forest fragments and *Eucalyptus* plantations. *Eucalyptus* plantations consisted of rodents (Atlantic Forest Climbing Mouse (*Rhipidomys mastacalis*), Black-footed Pygmy Rice Rat, Delicate Vesper Mouse, and Hairy-tailed Bolo Mouse (*Necromys lasiurus*)) and marsupials (White-eared Opossum (*Didelphis albiventris*) and Brazilian Gracile Opossum (*Gracilinanus microtarsus*)). Delicate Vesper Mouse was the only species occurring in all habitat types (semideciduous forests, *Eucalyptus*, *Pinus*, sugarcane plantations, and pastures).	Gheler-Costa et al. (2012)

Focus continent/ country	Species planted	Focal taxa	Major findings	Source
Brazil (Southeast)	*Eucalyptus* sp.	Birds	Species richness and abundance were significantly lower in eucalypt plantations compared to forest fragments and forest reserves. Only eight species were found in the plantations (Ruddy Ground Dove (*Columbina talpacoti*), Northern Slaty-Antshrike (*Thamnophilus punctatus*), Southern Beardless-Tyrannulet (*Camptostoma obsoletum*), Northern House Wren (*Troglodytes aedon*), Creamy-bellied Thrush (*Turdus amaurochalinus*), Red-eyed Vireo (*Vireo olivaceus*), Bananaquit (*Coereba flaveola*), and Palm Tanager (*Thraupis palmarum*)).	Marsden et al. (2001)
Brazil (Southwest)	*E. tereticornis*	Birds	Eucalypt (>70 years old) and native forest stands had similar species richness (36 and 32 species, respectively). Eucalypt stands did not significantly differ from eucalypt-native vegetation mosaics in terms of richness, abundance, species composition, and ecological aspects of understorey bird assemblages (i.e., diet, forest dependence, and sensitivity to forest fragmentation). Insectivores were the most common feeding guild (50%), followed by insectivores/frugivores (15.9%), granivores (9.1%), and frugivores and nectarivores (4.5% each).	Lopes et al. (2015)
Brazil (Southwest)	*Eucalyptus* spp.	Dung beetles	Eucalypt plantations had a significantly lower number of species compared to natural forests and grasslands. Plantations also harbored a subset of species commonly found in natural habitats. Similar communities were found in grasslands, natural candeial plantations, and 1-year-old eucalypt plantation. Four-year-old eucalypt plantations provided suitable habitat for forest specialists as indicator species of natural forest were found abundantly. Dung beetle communities prefer older eucalypt trees with higher canopy openness.	Gries et al. (2012)

Focus continent/ country	Species planted	Focal taxa	Major findings	Source
Brazil (Northeast)	*Eucalyptus* sp.	Birds	A low number of specialist species was found in eucalypt plantations compared to primary forests. Plantations recorded an abundance of nectarivores (Bananaquit and *Amazilia* spp.), the absence of arboreal granivores, and a low number of large arboreal frugivores and granivores.	Barlow et al. (2007b)
Brazil (Northeast)	*Eucalyptus* sp.	Mammals	Of the eight species recorded in eucalypt plantations, two were exclusively found in the plantations (i.e., Southern Tamandua *Tamandua tetradactyla* and Crab-eating Fox). Eucalypt plantations can support some mammalian diversity in the presence of natural habitats.	Piña et al. (2019)
Chile (Southwest)	*E. globulus*	Birds	Species richness did not differ between native forests (21 species) and abandoned eucalypt plantations (19 species), as they shared 90% of bird species. Mean number of individuals did not differ between the two sites. The most common species found in both sites were the Chucao Tapaculo (*Scelorchilus rubecula*), Green-backed Firecrown Hummingbird (*Sephanoides sephaniodes*), and White-crested Elaenia (*Elaenia albiceps*). Two species absent in abandoned eucalypt plantations were the Magellanic Woodpecker (*Campephilus magellanicus*) and Rufous-collared Sparrow (*Zonotrichia capensis*). Abandoned eucalypt plantations may provide suitable habitat for some native species due to the presence of native understorey vegetation. No nativeforest- dependent species were found in the plantations.	Fontúrbel et al. (2016)
Mexico (Northwest)	*E. globulus*	Spiders	Species richness in eucalypt plantations was higher than that of pine woodlands. Higher numbers of Lycosidae, Oxyopidae, and Salticidae species in eucalypt plantations, while Gnaphosidae and Corinnidae were more abundant in native woodlands. No habitat specialists were found. Spider species did not segregate by vegetation types.	Corcuera et al. (2010)

Focus continent/ country	Species planted	Focal taxa	Major findings	Source
Puerto Rico	*Eucalyptus* sp.	Common Coquí (*Eleutherodactylus coqui*)	There is no significant difference in adult frog density during the wet season in native forest and eucalypt plantations. Common Coquí density was influenced by vegetation structure more than plant species composition. Its density was not affected by plant species richness. Eucalypt plantations in montane areas provide suitable habitat for the species (limited to basal area thresholds of 18–28 m^2/ha).	Fogarty and Vilella (2003)
Portugal (Northwest)	*E. globulus*	Birds	A total of 28 species were found in eucalypt plantations, including 11 forest bird species. Forest bird species were lower in the plantations compared to oak and pine forests. Short-toed Treecreeper (*Certhia brachydactyla*), Eurasian Nuthatch (*Sitta europaea*), and Great Spotted Woodpecker (*Dendrocopos major*) were common species occurring in oak and/or pine patches but were absent in the plantations.	Proença et al. (2010)
Portugal (West)	*E. globulus*	Small mammals	Shrub cover affected the abundance of small mammals. High abundance of small mammals was found during the autumn season. Population structure was observed in White-toothed Shrew (*Crocidura russula*) species, where most females were captured in autumn and further away from the nearest plantation edge, while most males were captured in spring and closer to the plantation edge.	Carrilho et al. (2017)
Portugal (Central)	*E. globulus*	Macroinvertebrates	Larvae of Crane Fly (*Tipula lateralis*) preferred to consume alder leaves compared to eucalypt leaves. Larvae that fed on oak and eucalypt leaves did not grow well (survived for 63 and 91 days, respectively) compared to larvae that fed on alder leaves and chestnut (126 days). Deleterious changes in invertebrate communities and stream ecosystem processes may be affected by the replacement of *Eucalyptus* plantation in deciduous forests.	Canhoto and Graça (1995)

Focus continent/ country	Species planted	Focal taxa	Major findings	Source
Portugal (Central)	*E. globulus*	Macroinvertebrates	Monoculture eucalypts can lead to a lower (taxonomic) order of stream communities. Riparian corridors of original vegetation were important in reducing the impact of eucalypt plantations on invertebrate communities.	Abelho and Graca (1996)
Portugal (Central)	*E. globulus*	Macroinvertebrates	Riparian habitats had higher species abundance compared to 6–7-year-old eucalypt plantations due to higher soil moisture, with the exception of isopods, where detritivores were more abundant in the plantations due to the high amount of leaf litter. Low numbers of arthropods in the plantations may be due to the low density and diversity of herbs and shrubs, direct effects of leachates from eucalypt leaves, and the age of trees that supported abundant fauna.	Zahn et al. (2009)
Portugal (Central-east)	*Eucalyptus* sp.	Large mammals (Red Foxes (*Vulpes vulpes*), Stone Martens (*Martes foina*), Badger (*Meles meles*))	Negative correlation between Red Fox occupancy and plantations. No effect on interactions within the carnivore community, although Red Fox detectability increased with the presence of badgers. There was no evidence of temporal segregation among the three carnivores.	Cruz et al. (2015)
Portugal-Spain (Iberian Peninsula)	*Eucalyptus* sp.	Macroinvertebrates	Eucalypt leaves were poor food for shredders, but feeding rates increased when oils and polyphenolics were removed from the leaves. The effect of eucalypt plantations was not consistent on stream invertebrate communities. In Portugal, streams bordered by eucalypts had lower numbers of invertebrates compared to streams surrounded by deciduous forests, but the pattern was not clear in Spain. A short planting-to-harvest period of eucalypt plantations reduced the capacity to retain leaf litter and habitat availability to support invertebrates and fish.	Graça et al. (2002)

Focus continent/ country	Species planted	Focal taxa	Major findings	Source
Spain (North)	*Eucalyptus* sp.	Macroinvertebrates	Oligochaetes were found to be abundant in the plantations. Shredders (*Nemoura* spp. and *Echinogammarus* spp.) were significantly less abundant in the plantation compared to deciduous forests.	Larrañaga et al. (2006)
Spain (North)	*E. globulus*	Macroinvertebrates	Eucalypt plantations had significantly lower densities and proportions of adult shredder species (*Echinogammarus* spp.) compared to native deciduous forests. Shredder species (*Sericostoma pyrenaicum*) had similar densities in plantations and native deciduous sites, but the mass accumulation rate was lower in the plantations. The density of collector or grazer species (*Habroleptoides confusa*) was higher in the plantations compared to native deciduous sites.	Larrañaga et al. (2009)
Spain (North)	*E. globulus*	Macroinvertebrates	Eucalypt plantations did not show drastic impacts on stream community structure and function.	Pozo et al. (1997)
Spain (North)	*E. globulus*	Macroinvertebrates	The main taxa of macroinvertebrates (*Echinogammarus* sp., *Leuctra* spp., and *Nemouridae* spp.) were found more abundantly in alder litter compared to eucalypt litter. Both shredder and collector densities were higher in alder compared to eucalypt leaves upstream, but the difference in density disappeared downstream.	Basaguren and Pozo (1994)
Spain (Northwest)	*E. globulus*	Birds	Eucalypt plantations had the lowest richness (decreasing with age; young, intermediate, and mature) and abundance compared to pine plantations, native forests, and shrublands. They were a poor surrogate for native forest specialists, generalists, and wood habitat-associated species. Birds were associated with shrubs in young-age plantations and the canopy of older-age plantations. Certain bird species and feeding guilds changed according to plantation age. Limited phytophagous insects in plantations affected prey availability for birds.	Calviño--Cancela (2013)

Focus continent/ country	Species planted	Focal taxa	Major findings	Source
Spain (Northwest)	*E. globulus*	Macroinvertebrates	Macroinvertebrate diversity and richness decreased as eucalypt plantation cover increased. The trophic structure of streams changed from shredder dominance (caddisflies from Sericostomatidae and Brachycentridae) to collector dominance (fine organic matter detritivores, e.g., simuliids or limoniids) across native vegetation--eucalypt plantation cover.	Cordero--Rivera et al. (2017)
Spain (Northwest)	*Eucalyptus* sp.	Birds	Species richness and abundance of birds were higher in native forests compared to eucalypt plantations. A total of 34 bird species was recorded, with Eurasian Sparrowhawk (*Accipiter nisus*) exclusively found in the plantations.	Goded et al. (2019)
China (South)	*E. urophylla*, *E. grandis*	Fish	Mortality of Grass Carp (*Ctenopharyngodon idella*), Goldfish (*Carassius auratus*), and Eurasian Carp (*Cyprinus carpio*) in fishponds were likely caused by acute hypoxia and toxicity due to black water pollution from eucalypt plantation cutover.	Yang et al. (2019)
China (South)	*E. tereticornis*	Birds	Bird species richness and capture rate were lower in plantations compared to natural forests. *Eucalyptus* and *Pinus* plantations showed the greatest avifauna similarity compared to *Eucalyptus-Acacia* and *Eucalyptus-Schima*.	Zou et al. (2014)
Malaysia (Sabah)	*E. deglupta*	Macromoth	Species diversity in eucalypt plantations was as high as in natural secondary forests (872 and 1048, respectively) due to diverse and denser understorey plant species and habitat structural complexity. Noctuidae was the most occurring family in eucalypt plantations. *Cleora decisaria* was the most abundant species found in all plantation sites (i.e., *Acacia mangium*, *E. deglupta*, *Gmelina arborea*, *Paraserianthes falcataria*, and *Pinus caribaea*) as well as in the secondary forest.	Chey et al. (1997)

Focus continent/ country	Species planted	Focal taxa	Major findings	Source
Malaysia (Sabah)	*E. pellita*, *E. grandis*, *Eucalyptus* hybrid	Pest fauna	Two treeshrew species (i.e., Lesser Treeshrew *Tupaia minor* and Large Treeshrew) were found in *E. pellita* plantations, causing damage to tree barks. *Calliteara horsfieldii*, *Hyposidra picarian*, and *Darna trima* were eucalypt host moths in *Eucalyptus* hybrid plantations. Some pests in *E. pellita* and *E. grandis* were the Leaf-cutting Bee (*Megachile* sp.), micromoth larva (Tortricidae), and Gold Dust Weevil (*Hypomeces squamosus*). Termites of *Microcerotermes* spp. and *Macrotermes* spp. were found attacking *E. pellita*.	Chung et al. (2015)
Malaysia (Sabah)	*Eucalyptus* sp.	Birds	A survey reported species accounts, including the Little Bronze-Cuckoo (*Chrysococcyx minutillus*) in old tall trees of eucalypt groves.	Sheldon et al. (2009)
Malaysia (Sabah)	*E. deglupta*	Small mammals	Four types of plantations (*E. deglupta*, *Gmelina arborea*, *Albizia falcataria*, and *A. falcataria-Theobroma cacao*) were found to support scansorial small mammals of varying densities and abundances. Canopy and commensal species remained low in plantations, with Whitehead's Spiny Rat being the most abundant. A total of ten species were found in the plantations.	Stuebing and Gasis (1989)
Malaysia (Sabah)	*E. deglupta*	Large mammals	Mammal abundance was lower in plantations compared to natural forests, although the earlier mammalian colonizers (e.g., deer, Barking Deer (*Muntiacus muntjak*), civets, and Leopard Cats) were found in the plantations. A high abundance of mammals was present at the edge of a 6-year-old *E. deglupta-Gmelina* plantation and secondary forest.	Duff et al. (1984)
Malaysia (Sabah)	*Eucalyptus* sp.	Birds	A review concluded that eucalypt plantations were associated with lower plant and bird diversities. Bird composition was influenced by the age of eucalypt stands, which constitute different shrub structure, nesting, and pollination opportunities.	Zaiton et al. (2020)

Focus continent/ country	Species planted	Focal taxa	Major findings	Source
Malaysia (Pahang)	*Eucalyptus* sp.	Birds (montane)	Bird species richness in slightly disturbed forested areas (including a 2-ha plantation with *Acacia*, *Araucaria*, *Cupressus*, *Eucalyptus*, and *Pinus*) and pristine forest was similar, although the study pointed out the sensitivity of birds to habitat disturbance.	Soh et al. (2006)
Malaysia (Sarawak)	*Eucalyptus* sp.	Various wildlife	When initiating land preparation for new plantations, providing wildlife corridors should be an option by not clearing existing forests. Older plantations should be used to create wildlife corridors through replanting or restoring damaged buffers, especially along rivers.	Hon and Shibata (2013)
Sri Lanka	*Eucalyptus* spp.	Birds	A total of 17 species were found in all five habitats (home gardens, wetlands, tea plantations, eucalypt plantations, and small-scale reservoirs) with the Sri Lanka White-eye (*Zosterops ceylonensis*) being the most common. Raptors found in eucalypt plantations were the Black-winged Kite (*Elanus caeruleus*), Crested Serpent-eagle (*Spilornis cheela*), and Mountain Hawk-eagle (*Nisaetus nipalensis*). Species diversity was low in eucalypt plantations compared to other habitats, which could be due to a low abundance of insects.	Kottawa-Arachchi and Gamage (2015)
Australia	*E. globulus*, *E. nitens*	Various fauna	Eucalypt plantations provide habitat and resources for several bird species considered at risk, such as the Western Yellow Robin (*Eopsaltria griseogularis*), Inland Thornbill (*Acanthiza apicalis*), and Western Gerygone (*Gerygone fusca*). Bird and bat abundances were higher in plantations adjoining remnant vegetation compared to isolated plantations. Hollow-using arboreal marsupials were absent in plantations, while cavity-dependent birds were present but did not nest there.	Lindenmayer and Hobbs (2004)

Focus continent/ country	Species planted	Focal taxa	Major findings	Source
Australia (Southeast)	*Eucalyptus* sp.	Birds	The abundance of forest and woodland birds in eucalypt plantations was higher than in cleared farmland but lower than in native forests. Common insectivores foraging in the canopy and tall shrub layers were present in both native forests and plantations. Birds foraging on open ground were more common in plantations than in native forests. Nectarivores, carnivores, and birds foraging in low shrubs were less common in plantations compared to native forests.	Loyn et al. (2007)
Australia (Northwest Tasmania)	*Eucalyptus* sp.	Ground-dwelling invertebrates	Native land snail and millipede species diversities were lower in eucalypt plantations (16 species) compared to native forests (26 species). The abundance of introduced land snails and velvet worms were higher in eucalypt plantations.	Bonham et al. (2002)
Australia (North Tasmania)	*Eucalyptus* sp.	Birds	Mature eucalypts in plantations were important in maintaining biodiversity, especially for hollow-using species. Hollow use by birds was strongly related to hollow abundance in a tree. The most common hollow users were European Starling (*Sturnus vulgaris*) and Striated Pardalote (*Pardalotus striatus*).	Koch et al. (2009)
Australia (Southeast)	*E. pilularis*, *E. grandis*	Birds	Over 11 years of plantation growth, there was a 47% increase in bird species but a 31% decline in other species due to a matrix shift from cleared paddock to dense young eucalypt trees. The plantation matrix supported insectivores and nectarivores. Flowers from plantations influenced honeyeater abundance.	Law et al. (2014)

Focus continent/ country	Species planted	Focal taxa	Major findings	Source
Australia (Southeast)	*Eucalyptus* sp.	Birds	Eucalypt plantations provided significant improvement in bird population density compared to cleared or small tree paddocks. They were useful habitats for at least ten declining woodland-dependent species, including Speckled Warbler (*Pyrrholaemus sagittatus*), Red-Capped Robin (*Petroica goodenovii*), and Rufous Whistler (*Pachycephala rufiventris*). Similar bird communities can be found in mixed eucalypt-shrub plantations, remnant native forests, and woodlands.	Kavanagh et al. (2007)
Australia (Southeast)	*E. pilularis*	Birds	Eucalypt plantations with riparian strips had higher numbers of forest-dependent and habitat generalist species (25 and 23, respectively) compared to dryland eucalypt plantations (17 species). Riparian eucalypt plantations were characterized by five feeding guilds (e.g., omnivores and vertebrate eaters) compared to forest habitat with three feeding guilds (e.g., nectarivores and insectivores). Dryland eucalypt plantations had a significantly lower number of insectivore species than riparian plantations.	Hsu et al. (2010)
Australia (Southeast)	*Eucalyptus* sp.	Koalas	Occupancy of young plantations was strongly influenced by their proximity to remnant vegetation. Plantations have the potential to restore koala habitat.	Kavanagh and Stanton (2012)
Australia (West)	*E. globulus*	Insects	Leaf-feeding beetles had the highest abundance (79% of total specimens) in eucalypt plantations, with the Leaf-feeding Weevil (*Gonipterus scutellatus*) being the most abundant. Lower abundances of ants, bees, evanioid wasps, and predatory canopy beetles were found in the plantations compared to forest remnants and pastures. Species richness in eucalypt plantations was lower than that of forest remnants. Monoculture plantations supported a higher population of herbivore species.	Cunningham et al. (2008)

Focus continent/ country	Species planted	Focal taxa	Major findings	Source
Australia (Southwest)	*E. globulus*	Various fauna	Animals (e.g., reptiles, amphibians, and bats) used plantations less than the adjacent remnant native vegetation but more in open pasture. The abundance of animals (e.g., birds) was higher in remnant vegetation, especially at the edge areas.	Hobbs et al. (2003)
Africa (East)	*Eucalyptus* sp.	Birds	A total of 41 forest species were found in eucalypt plantations, with four species found exclusively there. Eucalypt plantations had lower species richness compared to natural forests, especially during spring migration and breeding periods. Well-managed eucalypt plantations can benefit some local species.	John and Kabigumila (2011)
Africa (South)	*E. nitens*	Cossid moth (*Coryphodema tristis*)	Native Cossid moths were found to undergo host shifts, infesting eucalypt trees (pest species attacking woody plants, including quince and vines). This species caused severe damage to *E. nitens* plantations but has yet to be found on other eucalypt species or in native woody plants in the *E. nitens* plantations.	Wingfield et al. (2008)
Madagascar	*Eucalyptus* spp.	Birds	Species richness of forest birds in eucalypt plantations was higher than other land uses (i.e., dry crops and rice paddies). Eucalypt plantations were identified as an important mosaic habitat for most feeding guilds.	Martin et al. (2012a)
Madagascar	*Eucalyptus* spp.	Madagascan Fruit Bat (*Eidolon dupreanum*)	Pollens and flower parts of eucalypt were found in the faeces of the endemic Madagascan Fruit Bat.	Picot et al. (2007)

References

Abelho M, Graca MAS (1996) Effects of eucalyptus afforestation on leaf litter dynamics and macroinvertebrate community structure of streams in Central Portugal. Hydrobiologia 324:195–204

Anjos N, Majer JD (2003) Leaf-eating beetles in Brazilian eucalypt plantations. Sch Environ Biol Bull 23:1–33

Barlow J, Gardner TA, Araujo IS et al (2007a) Quantifying the biodiversity value of tropical primary, secondary, and plantation forests. Proc Natl Acad Sci USA 104(47):18555–18560

Barlow J, Mestre LAM, Gardner TA et al (2007b) The value of primary, secondary and plantation forests for Amazonian birds. Biol Conserv 136(2):212–231. https://doi.org/10.1016/j.biocon.2006.11.021

Basaguren A, Pozo J (1994) Leaf litter processing of alder and eucalyptus in the Aguera stream system (northern Spain): II. Macroinvertebrates associated. Arch Hydrobiol 132(1):57–68

Bonham KJ, Mesibov R, Bashford R (2002) Diversity and abundance of some ground-dwelling invertebrates in plantation vs. native forests in Tasmania, Australia. For Ecol Manag 158(1–3):237–247. https://doi.org/10.1016/S0378-1127(00)00717-9

Calviño-Cancela M (2013) Effectiveness of eucalypt plantations as a surrogate habitat for birds. For Ecol Manag 310:692–699. https://doi.org/10.1016/j.foreco.2013.09.014

Canhoto C, Graça MAS (1995) Food value of introduced eucalypt leaves for a Mediterranean stream detritivore: *Tipula lateralis*. Freshw Biol 34(2):209–214. https://doi.org/10.1111/j.1365-2427.1995.tb00881.x

Cannell MGR (1999) Environmental impacts of forest monocultures: water use, acidification, wildlife conservation, and carbon storage. New For 17:239–262. https://doi.org/10.1023/A:1006551018221

Carrilho M, Teixeira D, Santos-Reis M et al (2017) Small mammal abundance in Mediterranean *Eucalyptus* plantations: how shrub cover can really make a difference. For Ecol Manag 391:256–263. https://doi.org/10.1016/j.foreco.2017.01.032

Chai DHS, Marzuki M, Kaliang H et al (2010) Incidence of stem borer in *Neolamarckia cadamba* (Kelampayan) plantation in Sarawak. In: Abstracts of the proceedings of International Symposium on Forestry and Forest Product (ISFFP), Forest Research Institute Malaysia

Chey VK, Holloway JD, Speight MR (1997) Diversity of moths in forest plantations and natural forests in Sabah. Bull Entomol Res 87(7):371–385. https://doi.org/10.1017/S000748530003738X

Chung AYC, Pang KNK, Kimjus K et al (2015) A note on some pests of *Eucalyptus* in Sabah, Malaysia. In: Abstracts of the proceedings of the IUFRO eucalypt conference 2015: scientific cultivation and green development to enhance the sustainability of eucalypt plantations, International Union of Forest Research Organizations

Corcuera P, Valverde PL, Zavala-Hurtado JA et al (2010) Non weaving spiders on native woodlands and *Eucalyptus* plantations in Western Mexico: diversity and distribution patterns. J Insect Conserv 14:711–719. https://doi.org/10.1007/s10841-010-9301-7

Cordero-Rivera A, Martínez Álvarez A, Cordero-Rivera A et al (2017) Eucalypt plantations reduce the diversity of macroinvertebrates in small forested streams. Anim Biodivers Conserv 40:87–97. https://doi.org/10.32800/abc.2017.40.0087

Cruz J, Sarmento P, White PCL (2015) Influence of exotic forest plantations on occupancy and co-occurrence patterns in a mediterranean carnivore guild. J Mammal 96(4):854–865. https://doi.org/10.1093/jmammal/gyv109

Cunningham SA, Floyd RB, Weir TA (2008) Do *Eucalyptus* plantations host an insect community similar to remnant *Eucalyptus* forest? Austral Ecol 30(1):103–117. https://doi.org/10.1111/j.1442-9993.2005.01429.x

Duff AB, Hall RA, Marsh CW (1984) A survey of wildlife in and around a commercial tree plantation in Sabah. Malays For 47(3–4):197–213

Feng Y, Ziegler AD, Elsen PR et al (2021) Upward expansion and acceleration of forest clearance in the mountains of Southeast Asia. Nat Sustain 4:892–899. https://doi.org/10.1038/s41893-021-00738-y

Flechtmann CAH, Ottati ALT, Berisford CW (2001) Ambrosia and bark beetles (Scolytidae: Coleoptera) in pine and eucalypt stands in southern Brazil. For Ecol Manag 142:183–191. https://doi.org/10.1016/S0378-1127(00)00349-2

Fogarty JH, Vilella FJ (2003) Use of native forest and *Eucalyptus* plantations by *Eleutherodactylus* frogs. J Wildl Manage 67(1):186–195. https://doi.org/10.2307/3803074

Fontúrbel FE, Candia AB, Castaño-Villa GJ (2016) Are abandoned eucalyptus plantations avifauna--friendly? A case study in the Valdivian rainforest. Rev Mex Biodivers 87(4):1402–1406. https://doi.org/10.1016/j.rmb.2016.09.011

Gheler-Costa C, Vettorazzi CA, Pardini R et al (2012) The distribution and abundance of small mammals in agroecosystems of southeastern Brazil. Mammalia 76(2):185–191. https://doi.org/10.1515/mammalia-2011-0109

Goded S, Ekroos J, Domínguez J et al (2019) Effects of eucalyptus plantations on avian and herb species richness and composition in North–West Spain. Glob Ecol Conserv 19:e00690. https://doi.org/10.1016/j.gecco.2019.e00690

Graça MAS, Pozo J, Canhoto C et al (2002) Effects of *Eucalyptus* plantations on detritus, decomposers, and detritivores in streams. Sci World J 2(1):1173–1185. https://doi.org/10.1100/tsw.2002.193

Gries R, Louzada J, Almeida S et al (2012) Evaluating the impacts and conservation value of exotic and native tree afforestation in Cerrado grasslands using dung beetles. Insect Conserv Divers 5(3):175–185. https://doi.org/10.1111/j.1752-4598.2011.00145.x

Hawes J, da Silva Motta C, Overal WL et al (2009) Diversity and composition of Amazonian moths in primary, secondary and plantation forests. J Trop Ecol 25(3):281–300. https://doi.org/10.1017/S0266467409006038

Hobbs R, Catling PC, Wombey JC et al (2003) Faunal use of bluegum (*Eucalyptus globulus*) plantations in southwestern Australia. Agrofor Syst 58:195–212. https://doi.org/10.1023/A:1026073906512

Hon J, Shibata S (2013) A review on land use in the Malaysian state of Sarawak, Borneo and recommendations for wildlife conservation inside production forest environment. Borneo J Resour Sci Technol 3(2):22–35. https://doi.org/10.33736/bjrst.244.2013

Hsu T, French K, Major R (2010) Avian assemblages in eucalypt forests, plantations and pastures in northern NSW, Australia. For Ecol Manage 260(6):1036–1046. https://doi.org/10.1016/j.foreco.2010.06.028

Intachat J, Chey VK, Holloway JD et al (1999) The impact of forest plantation development on the population and diversity of Geometrid Moths (Lepidoptera: Geometridae) in Malaysia. J Trop For Sci 11(2):329–336

Jamil NNM, Ibrahim H, Zain HHM et al (2020) Species diversity and feeding guilds of birds in Malaysian agarwood plantations. J Threat Taxa 12(14):16954–16961. https://doi.org/10.11609/jott.5213.12.14.16954-16961

John JRM, Kabigumila JD (2011) The use of bird species richness and abundance indices to assess the conservation value of exotic *Eucalyptus* plantations. Ostrich 82(1):27–37. https://doi.org/10.2989/00306525.2011.556799

Kavanagh RP, Stanton MA (2012) Koalas use young *Eucalyptus* plantations in an agricultural landscape on the Liverpool Plains, New South Wales. Ecol Manag Restor 13(3):297–305. https://doi.org/10.1111/emr.12005

Kavanagh RP, Stanton MA, Herring MW (2007) Eucalypt plantings on farms benefit woodland birds in South-Eastern Australia. Austral Ecol 32(6):635–650. https://doi.org/10.1111/j.1442-9993.2007.01746.x

Koch AJ, Munks SA, Spencer C (2009) Bird use of native trees retained in young eucalypt plantations: species richness and use of hollows. Wildlife Res 36(7):581–591. https://doi.org/10.1071/WR09037

Kottawa-Arachchi JD, Gamage RN (2015) Avifaunal diversity and bird community responses to man-made habitats in St. Coombs Tea Estate, Sri Lanka. J Threat Taxa 7(2):6878–6890. https://doi.org/10.11609/JoTT.o3483.6878-90

Larrañaga A, Larrañaga S, Basaguren A et al (2006) Assessing impact of eucalyptus plantations on benthic macroinvertebrate communities by a litter exclusion experiment. Ann Limnol 42(1):1–8. https://doi.org/10.1051/limn/2006002

Larrañaga A, Basaguren A, Pozo J (2009) Impacts of *Eucalyptus globulus* plantations on physiology and population densities of invertebrates inhabiting Iberian Atlantic streams. Int Rev Hydrobiol 94(4):497–511. https://doi.org/10.1002/iroh.200811156

Law BS, Chidel M, Brassil T et al (2014) Trends in bird diversity over 12 years in response to large-scale eucalypt plantation establishment: implications for extensive carbon plantings. For Ecol Manag 322:58–68. https://doi.org/10.1016/j.foreco.2014.02.032

Lindenmayer DB, Hobbs RJ (2004) Fauna conservation in Australian plantation forests–a review. Biol Conserv 119(2):151–168. https://doi.org/10.1016/j.biocon.2003.10.028

Lopes IT, Gussoni COA, Demarchi LO et al (2015) Diversity of understorey birds in old stands of native and *Eucalyptus* plantations. Restor Ecol 23(5):662–669. https://doi.org/10.1111/rec.12216

Loucks C, Ricketts TH, Naidoo R et al (2008) Explaining the global pattern of protected area coverage: relative importance of vertebrate biodiversity, human activities and agricultural suitability. J Biogeogr 35(8):1337–1348. https://doi.org/10.1111/j.1365-2699.2008.01899.x

Loyn RH, McNabb EG, Macak P et al (2007) Eucalypt plantations as habitat for birds on previously cleared farmland in southeastern Australia. Biol Conserv 137(4):533–548. https://doi.org/10.1016/j.biocon.2007.03.012

Marsden SJ, Whiffin M, Galetti M (2001) Bird diversity and abundance in forest fragments and *Eucalyptus* plantations around an Atlantic forest reserve, Brazil. Biodivers Conserv 10:737–751. https://doi.org/10.1023/A:1016669118956

Martin EA, Viano M, Ratsimisetra L et al (2012a) Maintenance of bird functional diversity in a traditional agroecosystem of Madagascar. Ecosyst Environ 149:1–9. https://doi.org/10.1016/j.agee.2011.12.005

Martin PS, Gheler-Costa C, Lopes PC et al (2012b) Terrestrial nonvolant small mammals in agro--silvicultural landscapes of Southeastern Brazil. For Ecol Manag 282:185–195. https://doi.org/10.1016/j.foreco.2012.07.002

Mitchell SL, Edwards DP, Martin RW et al (2022) Severity of deforestation mediates biotic homogenization in an Island archipelago. Ecography 2022(7):e05990. https://doi.org/10.1111/ecog.05990

Nair KSS (2001) Pest outbreaks in tropical forest plantations: is there a greater risk for exotic tree species? Center for International Forestry Research, Bogor

Omar H, Misman MA (2018) Time-series maps of aboveground biomass in dipterocarps forests of Malaysia from PALSAR and PALSAR-2 polarimetric data. Carbon Balance Manag 13:19. https://doi.org/10.1186/s13021-018-0108-2

Pellens R, Garayx I (1999) Edaphic macroarthropod communities in fast-growing plantations of *Eucalyptus grandis* Hill ex Maid (Myrtaceae) and *Acacia mangium* wild (Leguminosae) in Brazil. Eur J Soil Biol 35(2):77–89. https://doi.org/10.1016/S1164-5563(99)00209-5

Picot M, Jenkins RKB, Ramilijaona O et al (2007) The feeding ecology of *Eidolon dupreanum* (Pteropodidae) in eastern Madagascar. Afr J Ecol 45(4):645–650. https://doi.org/10.1111/j.1365-2028.2007.00788.x

Piña TEN, Carvalho WD, Rosalino LMC et al (2019) Drivers of mammal richness, diversity and occurrence in heterogeneous landscapes composed by plantation forests and natural environments. For Ecol Manag 449:117467. https://doi.org/10.1016/j.foreco.2019.117467

Poore MED, Fries C (1985) FAO forestry paper: the ecological effects of eucalyptus (vol. 59). Food and Agriculture Organization of the United Nations, Rome

Pozo J, González E, Díez JR et al (1997) Inputs of particulate organic matter to streams with different riparian vegetation. J N Amer Benthol Soc 16(3):602–611

Proença VM, Pereira HM, Guilherme J et al (2010) Plant and bird diversity in natural forests and in native and exotic plantations in NW Portugal. Acta Oecol 36(2):219–226. https://doi.org/10.1016/j.actao.2010.01.002

Punia H, Dhiman H, Saharan H et al (2022) Floristic composition and diversity in response to varying degrees of disturbance in tropical dry deciduous forests of Southern Haryana, India. Ecol Environ Conserv 28(4):2164–2173. https://doi.org/10.53550/EEC.2022.v28i04.079

Rahim A-KHA, Puan CL, Mah ANMMA et al (2023) Assessing cross-edge spillover of mammals in forest plantation through rapid camera-trapping. Malays Fores 86(1):49–62

Schabel HG, Hilje L, Nair KSS et al (1999) Economic entomology in tropical forest plantations: an update. J Trop For Sci 11(1):303–315

Sheldon FH, Styring AR (2011) Bird diversity differs between industrial tree plantations on Borneo: implications for conservation planning. Raffles Bull Zool 59(2):295–309

Sheldon FH, Lim HC, Nais J et al (2009) Observations on the ecology, distribution, and biogeography of forest birds in Sabah, Malaysia. Raffles Bull Zool 57(2):577–586

Shevade VS, Potapov PV, Harris NL et al (2017) Expansion of industrial plantations continues to threaten Malayan tiger habitat. Remote Sens 9(7):747. https://doi.org/10.3390/rs9070747

Sodhi NS, Koh LP, Brook BW et al (2004) Southeast Asian biodiversity: an impending disaster. Trends Ecol Evol 19(12):654–660

Soh MCK, Sodhi NS, Lim SLH (2006) High sensitivity of montane bird communities to habitat disturbance in peninsular Malaysia. Biol Conserv 129(2):149–166. https://doi.org/10.1016/j.biocon.2005.10.030

Stuebing RB, Gasis J (1989) A survey of small mammals within a Sabah tree plantation in Malaysia. J Trop Ecol 5(2):203–214. https://doi.org/10.1017/S0266467400003485

Styring AR, Ragai R, Unggang J et al (2011) Bird community assembly in Bornean industrial tree plantations: effects of forest age and structure. For Ecol Manag 261(3):531–544. https://doi.org/10.1016/j.foreco.2010.11.003

Suguituru SS, Silva RR, de Souza DR et al (2011) Ant community richness and composition across a gradient from Eucalyptus plantations to secondary Atlantic Forest. Biota Neotrop 11(1):369–376. https://doi.org/10.1590/S1676-06032011000100034

Sullivan BW, Nifong RL, Nasto MK et al (2019) Biogeochemical recuperation of lowland tropical forest during succession. Ecol 100(4):e02641. https://doi.org/10.1002/ecy.2641

Timo TPC, Lyra-Jorge MC, Gheler-Costa C et al (2015) Effect of the plantation age on the use of *Eucalyptus* stands by medium to large-sized wild mammals in southeastern Brazil. iForest 8(2):108–113. https://doi.org/10.3832/ifor1237-008

Wingfield MJ, Slippers B, Hurley BP et al (2008) Eucalypt pests and diseases: growing threats to plantation productivity. South For 70(2):139–144. https://doi.org/10.2989/SOUTH.FOR.2008.70.2.9.537

Wood TE, Cavaleri MA, Reed SC (2012) Tropical forest carbon balance in a warmer world: a critical review spanning microbial-to ecosystem-scale processes. Biol Rev 87(4):912–927. https://doi.org/10.1111/j.1469-185X.2012.00232.x

Yang G, Wen M, Deng Y et al (2019) Occurrence patterns of black water and its impact on fish in cutover areas of Eucalyptus plantations. Sci Total Environ 693:133393. https://doi.org/10.1016/j.scitotenv.2019.07.199

Zahn A, Rainho A, Rodrigues L et al (2009) Low macroarthropod abundance in exotic *Eucalyptus* plantations in the Mediterranean. Appl Ecol Environ Res 7(4):297–301

Zaiton S, Paridah MT, Hazandy AH et al (2018) Potential of *Eucalyptus* plantation in Malaysia. Malays Fores 81(1):64–72

Zaiton S, Sheriza MR, Ainishifaa R et al (2020) Eucalyptus in Malaysia: review on environmental impacts. J Landsc Ecol 13(2):79–94. https://doi.org/10.2478/jlecol-2020-0011

Zanzini ACS, Silva AAN, Pereira CZ, Santiago WTV, Zanon MHC (2017) Composition of species of medium and large sized mammals in *Eucalyptus* plantations and native forests fragments, in the southeast of Minas Gerais state, Brazil. Sci For 45(116):673–683. https://doi.org/10.18671/scifor.v45n116.08

Zeng Z, Estes L, Ziegler AD et al (2018) Highland cropland expansion and forest loss in Southeast Asia in the twenty-first century. Nat Geosci 11:556–562. https://doi.org/10.1038/s41561-018-0166-9

Zou F, Yang Q, Lin Y et al (2014) Bird community comparisons of four plantations and conservation concerns in South China. Integr Zool 9(1):97–106. https://doi.org/10.1111/1749-4877.12037

A Review of *Eucalyptus*, Microclimate, and Forest Fire Relations

Ruzana Adibah Mohd Sanusi, Ahmad Ainuddin Nuruddin, and Haida Zainol

Abstract The conversion of the natural environment to forest plantations alters the surrounding environment and vegetation structure and consequently modifies the microclimate conditions. These changes in land use and microclimate can influence rainfall patterns and increase the likelihood of fire occurrence. Forest plantations are particularly susceptible to fire due to their vegetation structure, which often includes grasses, bushes, and highly flammable tree species such as *Eucalyptus*. This review aims to investigate the extent of and advances in existing research on the relationship between *Eucalyptus*, microclimate changes, and forest fire occurrence. A comprehensive review of relevant literature was conducted, synthesizing past studies on selected topics and assessing the scope and depth of information presented in the research. Key findings from the reviewed articles are discussed to identify prevailing themes and gaps. This review concludes by providing recommendations for future studies, such as furthering research on the physiological performance, biodiversity, hydrology, and soil processes.

1 Introduction

Land changes and modifications from natural areas to planted forests will influence the landscape characteristics. Land cover change is a major driver of climate change and, at the microscale, leads to modifications in the local microclimate. These changes, in turn, affect various ecosystem services, including flora and fauna

R. A. Mohd Sanusi (✉)
Institute of Tropical Forestry and Forest Products, Universiti Putra Malaysia, Serdang, Selangor, Malaysia

Faculty of Forestry and Environment, Universiti Putra Malaysia, Serdang, Selangor, Malaysia
e-mail: ruzanasanusi@upm.edu.my

A. A. Nuruddin · H. Zainol
Institute of Tropical Forestry and Forest Products, Universiti Putra Malaysia, Serdang, Selangor, Malaysia

S. Mohd Razali et al. (eds.), *Sustainable Eucalyptus Forestry in Southeast Asia*, Forestry Sciences 88, https://doi.org/10.1007/978-3-032-21343-3_3

biodiversity, soil carbon dynamics, and other ecological functions (Williams and Newbold 2020; Lozano-García et al. 2017; Fugère et al. 2016).

One of the primary forces behind deforestation is the rising global demand for timber. The increasing pressure to produce wood has accelerated the establishment of forest plantations. Globally, the area of forest plantations increased from 167.5 to 277.9 million ha between 1995 and 2015, with the observed increase varying by regional and climatic conditions (Payn et al. 2015). At the same time, the estimation increased by 294 million ha of the global planted forest area in 2020 (FAO 2020). On the other hand, projections by Korhonen et al. (2021) estimated that the planted forest is expected to increase up to 334 million ha, representing a 14% increase compared to 293 million ha in 2015. Planted forests should be seen as a complement rather than an alternative to natural forests (Evans 2009), as they offer a means to improve ecosystem services in degraded areas and simultaneously support the wood-based industries.

Fast-growing species, such as pines and eucalypts, are typically selected for rapid timber production; however, in many instances, they are usually planted outside of their native setting (Wyse et al. 2022; Cordero Rivera 2017). According to Fernandes et al. (2016), fast-growing and highly flammable trees such as pines and eucalypts pose a significant fire threat to the surroundings, potentially leading to high-intensity and large-scale fires. Moreover, the effects of climate change, coupled with microclimate shifts towards hotter and drier conditions, can alter rainfall patterns and increase the frequency of fire occurrences (Lai and Zhang 2025; Mann et al. 2016).

Given the detrimental effects of changes in the landscape and microclimate conditions, as well as plantation forest fires, on both the environment and society, it is crucial to establish knowledge based on the extent to which microclimate studies on Eucalyptus plantations contribute to the broader study of forest fires. Therefore, this chapter aims to review the existing literature on the microclimate effects of Eucalyptus plantations and their relationship with forest fire dynamics. Specifically, this review seeks to highlight the importance of this issue and assess the current scope of research. It is guided by the following specific research questions:

1. What research has been conducted on microclimate and Eucalyptus plantations? Are there any forest fire-related studies associated with microclimate conditions?
2. What types of information and methodologies have been used in these studies?

This review will examine key perspectives on microclimate and fire risk in Eucalyptus plantations and then conclude with recommendations for future studies to further advance knowledge in this field.

2 Methodology

This literature search was divided into two parts, where the first part focused on the specific research done in relation to microclimate and Eucalyptus plantation. This was then followed by a literature investigation on Eucalyptus plantation microclimate conditions and their association with forest fire. The search string was composed of this perspective, and the search was applied to the "Titles, Abstracts and Keywords" in the Scopus database.

The first set of search strings was related to microclimate and Eucalyptus plantation with: (TITLE-ABS-KEY (microclimate) AND TITLE-ABS-KEY (eucalyptus) AND TITLE-ABS-KEY (plantation)). While the second set of search strings was on Eucalyptus plantation, microclimate condition and forest fire with: (TITLE-ABS-KEY (microclimate) OR TITLE-ABS-KEY (climate) AND TITLE-ABS-KEY (plantation) AND TITLE-ABS-KEY (eucalyptus) AND TITLE-ABS-KEY (fire) OR TITLE-ABS-KEY ("forest fire")).

All the abstracts of the returned articles were checked to make sure that they are within the research scope. Data gathered from the returned articles were read based on each search string, where the information on number of articles, studied species, topic of study, year of publication, scale of research and data collection process were obtained. The findings from each study were also summarized, and the similarities and differences between the returned studies were further discussed.

3 Results and Discussion

3.1 Returned Articles

From the literature searches, for the first set of search strings on microclimate and Eucalyptus (Fig. 1), the performed search returned a total of 28, including journal articles, books, and book series. While for the second set of search strings in relation to forest fire, this search returned 20 papers, including only journal articles, book chapters, and conference papers (Fig. 2). For the first set of literature search, 8 papers were excluded, while 7 papers were excluded for the second set, as there was no clear relation to the investigated subjects, review articles, one article was redundant in both searches (only included in the search related to forest fire) and multiple articles only discussed but were not relevant as the main research subject.

Both searches only returned one redundant article from Wright et al. (2012), while all other articles were unique to their respective searches. This outcome indicates the importance of conducting separate searches, as studies on microclimate and forest fires in Eucalyptus plantations, although seemingly interrelated, are seldom investigated in tandem. It also indicates the need for integrated studies that explicitly examine the relationship between microclimate dynamics and fire risk in Eucalyptus plantations. Moreover, with only 48 articles returned in total, this explains the limited research attention on the intersection of these two critical elements, despite their relevance to ecological sustainability and the resilience of Eucalyptus plantation systems.

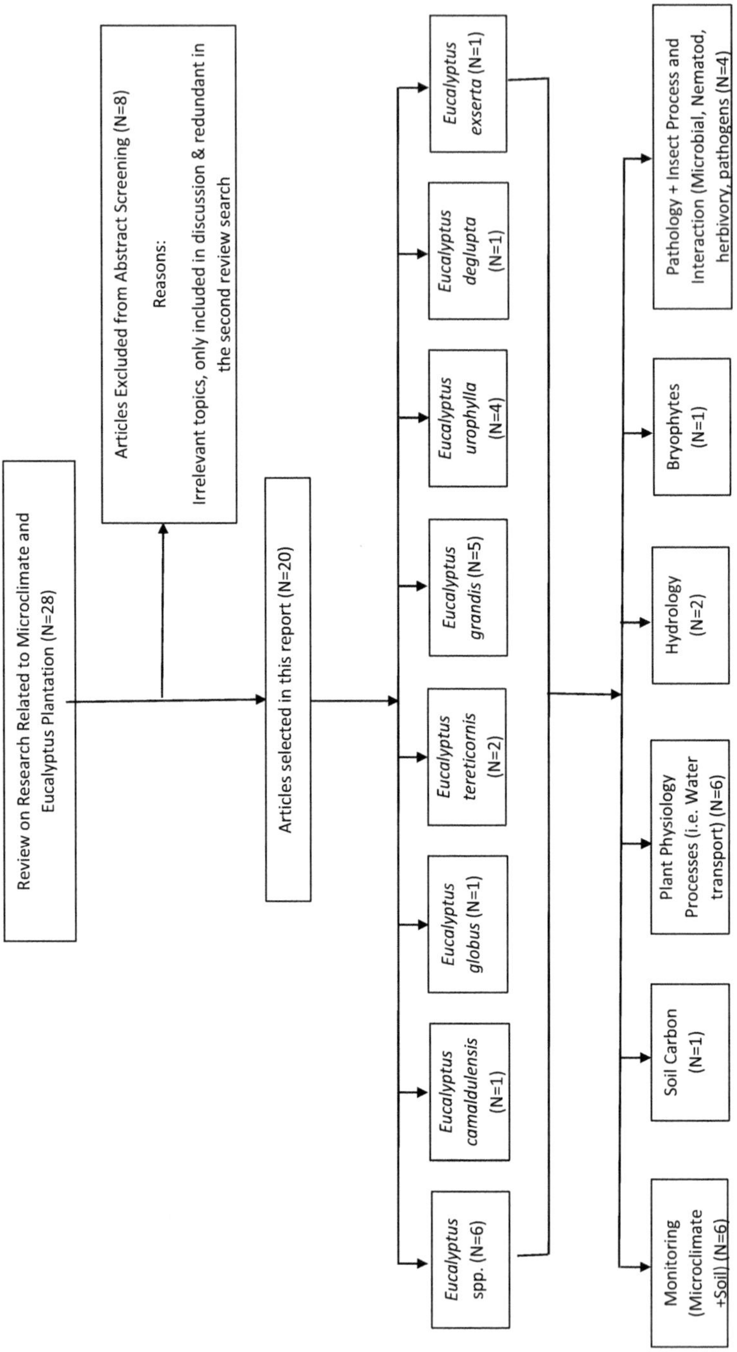

Fig. 1 Flow diagram from the first set of search strings on articles related to microclimate and Eucalyptus plantations

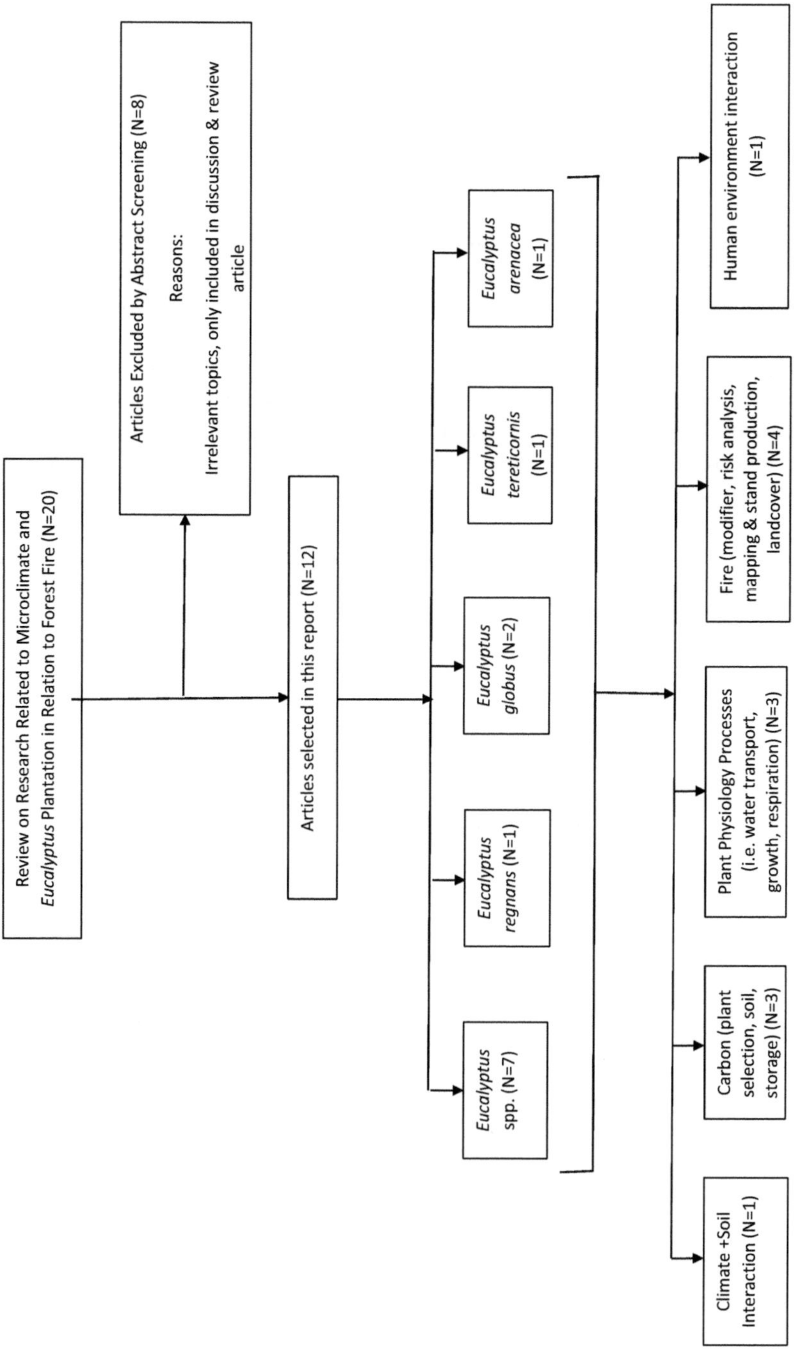

Fig. 2 Flow diagram from the second set of search strings on articles related to microclimate and Eucalyptus plantation in relation to forest fires

3.2 Studied Species

The search for microclimate and Eucalyptus plantations yielded studies involving various species of Eucalyptus. Six studies did not mention the specific species of Eucalyptus (Fig. 1). Among the identified species, *E. grandis* was the most frequently studied ($N = 5$), followed by *E. urophylla* ($N = 4$) and *E. tereticornis* ($N = 2$). Other species, namely *E. camaldulensis*, *E. exserta*, *E. deglupta*, and *E. globulus*, were each investigated in only one study. In comparison, the search related to forest fires had fewer returned articles, resulting in only five species of Eucalyptus (Fig. 2). The most common was *Eucalyptus* spp. (unstated species) ($N = 8$), followed by *E. globulus* ($N = 2$), while *E. tereticornis*, *E. arenacea* and *E. regnans*, each represented one study conducted for each species.

Overall, it is interesting to see various species of Eucalyptus used in the reviewed studies. However, 40.6% of the studies ($N = 13$) did not clearly specify the species examined, making it difficult to actually determine which species are most frequently investigated globally. Nonetheless, *E. grandis* (15.6%) and *E. urophylla* (12%) emerged as species of relatively high research interest, indicating their significance, particularly in relation to microclimate studies. Meanwhile, other species, including *E. globulus*, *E. camaldulensis*, *E. exserta*, *E. deglupta*, *E. tereticornis*, *E. arenacea*, and *E. regnans* were limitedly used as subject species. It is interesting to note that although some species are widely planted and distributed, such as *E. camaldulensis* and *E. globulus*, studies examining their interactions with microclimate and forest fire dynamics remain limited.

3.3 Topics of Research

There were several overlapping research themes across both sets of search results. Overall, the identified studies focused on five to six primary topics, including the surrounding climate, plant physiological processes, microclimate and soil monitoring, and forest fire analysis. Specifically, in microclimate studies, research on Eucalyptus predominantly focused on microclimate and soil conditions ($N = 6$), plant physiology processes, particularly water transport ($N = 6$), and interactions involving plant pathology and insect dynamics, such as microbial, nematode, herbivory, and pathogen studies ($N = 4$). In contrast, topics such as hydrological processes ($N = 2$), bryophytes ($N = 1$), and soil carbon ($N = 1$) received comparatively less attention.

For studies related to forest fires, most studies focused on fire patterns, including fire as a modifier, risk analysis, mapping stand production, and the influence of land cover on fire behaviour ($N = 4$). This was followed by studies on plant physiological processes, specifically water transport, growth, and respiration ($N = 3$), as well as carbon-related topics, including plant selection, soil carbon, and carbon storage

(*N* = 3). Meanwhile, topics exploring the human–environment interactions (*N* = 1) and climate–soil interactions (*N* = 1) were the least represented.

The topics studied within the Eucalyptus plantation concerning microclimate and forest fires were specific, with plant physiological processes, microclimate, soil monitoring, and forest fire analysis emerging as the dominant themes. It highlights the relevance of these topics to Eucalyptus plantations, where land conversion from natural forest to forest production lands and the selection of Eucalyptus species for plantation and degraded area restoration are the core interests. Furthermore, the emphasis on plant physiological responses of Eucalyptus indicates the importance of this topic in understanding microclimate and forest fires.

3.4 Year of Publication

The selected returned articles were published between 1985 and 2020. The publication years were grouped into six categories: 2020–2015, 2014–2010, 2009–2005, 2004–2000, 1997, and 1989 (Table 1). The years 1997 and 1989 were not grouped into ranges, as only a small number of studies were published in those specific years. For the first search string, which concerned microclimate and Eucalyptus plantations, the publication span covered 46 years. The most active period was between 2014 and 2010 (*N* = 7), followed by 2020–2015 (*N* = 4), 2004–2000 (*N* = 3), 2009–2005 (*N* = 2), 1997 (*N* = 2), and one article each from 1989 and 1974

Table 1 Components assessed in all selected studies, including year of studies, study scale, and data collection process for both search strings

Information assessed/ Level	Category	First search string (eucalyptus & microclimate)	Percentage (%)	Second search string (eucalyptus, microclimate, & fires)	Percentage (%)
Year of studies	2020–2015	4	20	7	58.3
	2014–2010	7	35	3	25.0
	2009–2005	2	10	2	16.7
	2004–2000	3	15	–	–
	1997	2	10	–	–
	1989	1	5	–	–
	1974	1	5		
Scale	Site	20	100	3	25.0
	Regional	0	0	8	66.7
	Multi countries	0	0	1	8.3
Data collection process	Field-based	20	100	3	25.0
	Process/ Mechanistic modeling	0	0	9	75.0

($N = 1$). The second search string, which concerned forest fires, only yielded publications between 2008 and 2020. The highest number of publications occurred during 2020–2015 ($H = 7$), followed by 2014–2010 ($N = 3$) and 2009–2005 ($N = 2$).

These patterns suggest that research on Eucalyptus and microclimate has been ongoing for several decades. However, studies explicitly exploring the triadic relationship between Eucalyptus, microclimate, and forest fires only emerged from 2008 onward. This more recent trend suggests a growing recognition of the importance of examining how Eucalyptus plantations interact with changing climatic conditions and fire dynamics. The evolving focus of research highlights an opportunity for methodological innovation. Future studies may benefit from incorporating modern technologies, such as real-time monitoring using Internet of Things (IoT) systems, to capture dynamic environmental variables.

3.5 Scale of Research and Data Collection Approach

In terms of the research scale, the two search categories revealed contrasting patterns. All studies identified in the first search, which focused on microclimate effects from Eucalyptus stands, were site-specific in nature (Table 1). Conversely, studies from the second search, which focused on forest fires, were more diverse in scale, where 66.7% ($N = 8$) of the studies were done at a regional scale, while 25.0% ($N = 3$) were site-specific and 8.3% ($N = 1$) involved cross-country or multinational comparisons (Table 1). When combining results from both searches, the overall distribution showed that 65.6% of the studies were site-specific, 25% were regional, and 3.1% were multinational (Fig. 3a).

Regarding data collection methods, all studies in the microclimate and Eucalyptus stands category employed field-based approaches (100%), whereas 75% of forest fire-related studies relied on process-based and mechanistic modelling approaches (Table 1). When considered together, 71.9% of all studies employed field-based methods, while the remaining 28.1% used process-based and mechanistic modeling approaches (Fig. 3b).

The variations in research scale and data collection methods reflect the distinct characteristics of microclimate and forest fire studies. The site-specific approach is logical for microclimate studies, which operate at a micro level and necessitate precise site-specific data for accurate results. In contrast, forest fire studies often rely on large datasets to predict fire behaviour, patterns, and stand interactions with climate, soil, human factors, and carbon dynamics. Interestingly, a subset of forest fire studies, particularly those examining plant physiological responses, were conducted at the site level because they require direct measurements of parameters such as plant water transport, growth, and respiration. These physiological indicators are essential for understanding how fire affects, and is affected by, Eucalyptus stand conditions.

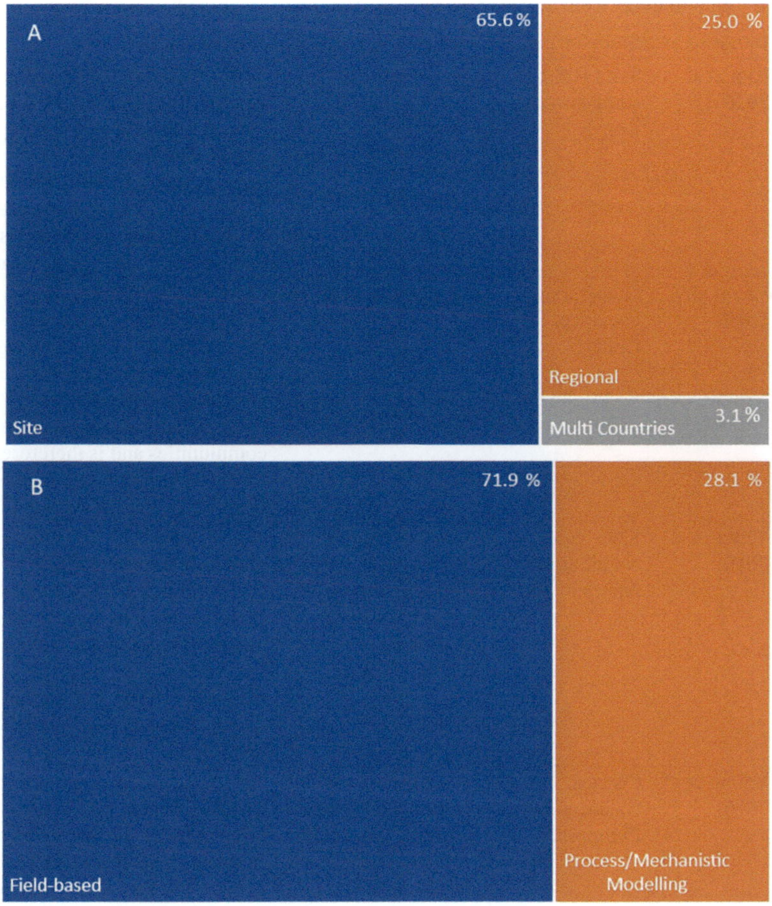

Fig. 3 Overall percentage of returned articles related to microclimate as well as its influence on forest fire on (**a**) scale and (**b**) data collection approach

3.6 Findings from the Returned Articles

The abstracts of the returned articles were further investigated to identify key findings relevant to the study objectives. The main findings of the reviewed literature are summarized in Tables 2 and 3.

The canopy cover of Eucalyptus plays an important role in regulating microclimate conditions. For instance, a 20% reduction in leaf area index (LAI) can result in a 1.2 °C increase in ground temperature (Wan et al. 2019). However, Lemma et al. (2007) reported in a study that the LAI of *E. grandis* was the lowest compared to that of *Cupressus lusitanica* and *Pinus patula,* potentially influencing the within- and below-canopy microclimates. This highlights the importance of species selection in plantation planning and suggests that microclimatic effects should be

Table 2 Main findings from articles related to microclimate and Eucalyptus plantation

No	Author	Study origin	Species	Main findings
1	Stam et al. (2020)	Taita Hills, southeastern Kenya	*Eucalyptus* spp.	Transplant growth showed no significant difference between natural locations and Eucalyptus plantations, indicating that the primary factor limiting bryophyte biomass was the lack of suitable substrate in plantation forests rather than an unsuitable microclimate.
2	Wan et al. (2019)	Guangdong Province, China	*E. urophylla*	The removal of understory vegetation affected the soil microclimate and reduced the soil's dissolved carbon content. Understory vegetation helps maintain soil microbial communities and is therefore recommended not to be removed from Eucalyptus plantations.
3	Pfeifer et al. (2019)	Kwa-Zulua Natal, South Africa	Bushland, natural forests, edges of natural forests, and timber plantations comprising *Eucalyptus* spp.	A 20% forest gap or a one-unit reduction in canopy leaf area index (LAI) led to a 1.2 °C increase in ground temperature. Microscale temperature variation is important for improving biodiversity. However, further mapping of the demographic traits of species is needed.
4	Paelmo et al. (2016)	Cuenca, Batangas, Philippines	*Jatropha curcas* L. planted in SRF-based agroforestry system (*Acacia mangium, Pongamia pinnata,* and *Eucalyptus deglupta*)	Air and soil temperature, along with wind velocity, were significantly influenced by the Short Rotation Forestry (SRF)-based agroforestry system, but not relative humidity. Only SRF-based agroforestry systems using *A. mangium* and *E. deglupta* showed beneficial effects on site conditions.
5	Zhao et al. (2014)	Guangdong Province of China	*E. urophylla*	Soil microbial and nematode diversities were affected by the overstory and understory vegetation. Soil microclimate modification due to vegetation removal was the determining factor in the variance of soil biodiversity indices.

(continued)

Table 2 (continued)

No	Author	Study origin	Species	Main findings
6	Wang et al. (2014)	Southern China	*E. urophylla* and *Acacia crassicarpa*	After 6 months, soil temperature was reduced by 0.5 °C–0.8 °C following understory removal (UR) in *E. urophylla* and *A. crassicarpa* plantations, while soil moisture decreased in *E. urophylla*. Soil organic matter also declined due to understory vegetation removal.
7	Zhao et al. (2013)	Guangdong Province of China	*E. urophylla*	Soil temperature increased due to understory vegetation removal. However, it also reduced soil water content, fungal biomass, and fungal-to-bacterial biomass. In addition, soil temperature, soil water content, and soil microbial community were not significantly affected by nitrogen fertilization. This indicates the predominant influence of understory vegetation on soil microclimate and soil microbial communities in Eucalyptus plantations.
8	Tjoelker et al. (2013)	Sydney, Australia	*E. grandis*	Sap flux patterns increased with increasing tree height along the stem axis but not in the branches. All trees showed no signs of water stress and adequate transpiration rates.
9	Liu et al. (2012)	Guangdong Province of China	*Eucalyptus* spp.	Soil moisture and temperature influenced microbial metabolic activity and substrate utilization patterns. Microbial metabolic activity for all carbon sources and the specific substrate metabolic activities were positively correlated with soil temperature. Metabolic diversity and evenness were positively correlated with soil temperature; however, soil moisture was negatively correlated with richness.

(continued)

Table 2 (continued)

No	Author	Study origin	Species	Main findings
10	Arnold and Fonseca (2011)	Southern Brazil	Araucaria forest and ecologically managed plantations of Araucaria, Pinus, and Eucalyptus	No significant differences in the mean and frequency distribution pattern of leaf damage between the *Araucaria* forest and the three monoculture plantations. Due to the shadier and moister microclimate, Eucalyptus plantations had a higher percentage of leaves with epiphylls than the *Araucaria* Forest.
11	do Nascimento (2010)	São Paulo, Brazil	*Eucalyptus* spp.	The use of Eucalyptus as a barrier lowered the light intensity by 35%, lowered air temperature by 1 °C, and increased relative humidity by 3.4% throughout the year. It suggests that Eucalyptus stands along the edge of forest fragments are effective for microclimate mitigation.
12	Kallarackal and Somen (2008)	Peninsular India	*E. tereticornis*, *E. grandis*, *Acacia auriculiformis*, *Anacardium occidentale,* and *Tectona grandis*	Transpiration varied across species due to the differences in rooting depth, leaf area index, stomatal conductance, and soil water availability. These ecophysiological traits, along with water loss during transpiration, should inform species selection for tropical plantation development.
13	Lemma et al. (2007)	Southwestern highlands of Ethiopia	*E. grandis, Cupressus lusitanica* and *Pinus patula*	The Eucalyptus stand had the lowest LAI with 2.07 ± 0.05 $m^2\ m^{-2}$ compared to other species, potentially affecting the microclimate both within and below the canopy. However, soil organic carbon (SOC) storage was primarily influenced by total litter input and the proportion of fine woody litter (branches and coarse roots), not litter quality or microclimate.

(continued)

Table 2 (continued)

No	Author	Study origin	Species	Main findings
14	Zhou and Yan (2003)	Southern China	*Eucalyptus* spp.	Rainfall interception and surface runoff were higher in the mixed forest canopy than in Eucalyptus plantations, with lower stemflow. Bare land experienced 5 and 129 times more soil erosion than Eucalyptus and mixed forests, respectively. Eucalyptus and mixed forests had deeper water tables—80 cm and 30 cm deeper, respectively—than bare land, indicating the potential of Eucalyptus for the rehabilitation of degraded lands.
15	Zhou et al. (2002)	Dianbai County, Guangdong Province, China	*E. exserta*	Eucalyptus plantations and mixed forests had higher absolute humidity and lower maximum near-ground temperatures compared to bare land. Water runoff was reduced in *Eucalyptus* catchments due to accumulated litter and understorey vegetation.
16	Mielke et al. (2000)	Espirito Santo, Brazil	*E. grandis* Hill ex. Maiden hybrids	High soil water availability contributed to high stomatal conductance (g_s) and net photosynthetic rate (A) with values over 0.4 mol m^{-2} s^{-1} and 15 μmol m^{-2} s^{-1}, respectively. The main microclimatic variables controlling g_s were mainly the photosynthetic active radiation (Q), vapor pressure deficit (VPD), atmospheric CO_2 molar fraction (C(a)), and leaf water potential Δ(pd), while A was affected by Q and VPD.
17	Kallarackal and Somen (1997a)	Kerala State Forest in peninsular India	*E. grandis*	*E. grandis* had good stomatal control, thus more effective regulation of transpiration, especially during dry seasons. Dry conditions had minimal effects on the trees' photosynthetic rate.

(continued)

Table 2 (continued)

No	Author	Study origin	Species	Main findings
18	Kallarackal and Somen (1997b)	Kerala State Forest, India	*E. tereticornis* Sm.	Stomatal conductance was strongly correlated with atmospheric vapour pressure deficit; however, even at high atmospheric VPD (≃ 5.0 kPa), stomata did not completely close. During the pre-monsoon period, tree growth was not significant, and mild water stress did not affect photosynthesis.
19	Pereira et al. (1989)	Alentejo, Portugal	*E. globulus* and *Quercus rotundifolia*	There was no significant difference in evapotranspiration between native and non-native species. However, Eucalyptus harvesting increased runoff by 5 cm year^{-1} relative to the undisturbed oak forests.
20	Schiller and Karschon (1974)	Be'er Sheva'-Dimona	*E. camaldulensis* Dehn.	Air temperature and vapour pressure did not differ significantly between the flood-irrigated *E. camaldulensis* plots and open desert areas. However, black-body temperature, global radiation, evaporation, and wind speed were lower in *E. camaldulensis* stands, reducing thermal and providing recreational opportunities to the public. However, winter temperatures in the *E. camaldulensis* stands limit recreational activities.

investigated before selection. Eucalyptus plantations and other forest cover in general provide greater absolute humidity and lower maximum near-ground temperatures compared to bare land (Wright and Francia 2024; Zhou et al. 2002). In addition, planting Eucalyptus along the edges of forest fragments has been shown to moderate microclimatic conditions, as it lowered the light intensity by 35%, lowered air temperature by 1 °C, and increased relative humidity by 3.4% throughout the year (do Nascimento 2010).

Multiple articles also demonstrated that the removal of understory vegetation significantly alters local environmental conditions. For instance, the removal of understory vegetation under Eucalyptus stands impacts air and soil microclimate, and reduces soil dissolved carbon contents, soil organic matter, microbial activities, and nematode diversities (Wan et al. 2019; Zhao et al. 2013, 2014; Wang et al. 2014; Liu et al. 2012). In addition, Eucalyptus catchments reduce water runoff due to the accumulation of litter and understory vegetation (Castagnolli et al. 2023; Zhou et al. 2002). In another finding, litter quality and microclimate had a limited contribution to differences in soil organic carbon storage (SOC) for *E. grandis*, *C. lusitanica*, and

Table 3 Main findings from articles related to microclimate and Eucalyptus plantation in relation to forest fires

NO	Author	Location	Species	Main findings
1	Buchanan and Kennedy (2009)	Central Victoria, Australia	*E. globulus* and *Pinus radiata*	*E. globulus* was projected as the preferred species over *P. radiata* based on climate-change scenario analysis. The analysis serves as an important tool in management, particularly for plant selection in combating the challenges of climate change
2	Elli et al. (2019)	Brazil	*Eucalyptus* spp.	Growth gaps in Eucalyptus plantations were significantly affected by water deficit, with growth reductions of 87% and 67% reported in the Southeast and North regions, respectively. Recommended strategies include the use of drought-tolerant genotypes, wider spacing with deeper subsoiling, improved soil management, along with monitoring of forest fires, windstorms, and waterlogging.
3	Fernandes et al. (2019)	Portugal	*Eucalyptus* spp.	The expansion of Eucalyptus had no significant impact on increasing burned areas in Portugal. Forest composition had little influence on fire size and fuel hazard.
4	Bowman et al. (2019)	Central Chile	*Eucalyptus* spp.	The warm and moist weather conditions further intensified fire activity in 2016, following an intense drought that started in 2010. Approximately 80% of the land has been converted to Pinus and Eucalyptus plantations since the 1970s. In 2017, 44% of these plantations were burned, marking some of the most severe fires in three decades. This highlights the vulnerability of such plantations to fire incidents, with increasing risk of more extreme wildfire seasons.

(continued)

Table 3 (continued)

NO	Author	Location	Species	Main findings
5	Alegria and Pedro (2019)	Central Inland Portugal	*Eucalyptus* spp.	The total area of Eucalyptus forests did not significantly increase; however, fire occurrences in pine forests have led to Eucalyptus afforestation. Therefore, implementing sound management practices in Eucalyptus plantations is essential to prevent soil fertility loss due to nutrient deficiencies, organic matter depletion, or erosion.
6	Baqir et al. (2019)	Mau, Uttar Pradesh, India.	Multiple species, including *Prosopis juliflora, Acacia nilotica, Eucalyptus* spp., *Terminalia arjuna, Tectona grandis*	Among the 10 dominant tree species studied, *P. juliflora* (52.0%), *A. nilotica* (48.5%), *T. grandis* (46.4%), and *Eucalyptus* spp. (45.4%) showed strong carbon sequestration potential. Fuelwood consumption varies depending on the season. Based on calorific value, *P. juliflora*, *T. arjuna*, *Eucalyptus* spp., and *T. grandis* showed potential as tree species for energy plantations.
7	Arora and Chaudhry (2017)	Kurukshetra, India	*Acacia nilotica* + *Dalbergia sissoo*, *Syzygium cumini*, *Tectona grandis*, *Populus deltoides* and *E. tereticornis*	*A. nilotica* + *D. sissoo* and *E. tereticornis* had significant correlations with soil respiration rate and soil temperature. Significant correlations between soil respiration rate and rainfall were observed for *S. cumini*, *T. grandis*, *P. deltoides* and *E. tereticornis*. Soil moisture and temperature had an effect on microbial activity, increasing the soil organic matter decomposition rate and soil CO_2 efflux. Forest fires and land-use changes can alter these soil processes.

(continued)

Table 3 (continued)

NO	Author	Location	Species	Main findings
8	Jactel et al. (2012)	Portugal, France, UK (Scotland), Poland, Sweden, Germany, Austria	Five tree species (*E. globulus*, *Pinus pinaster*, *Pinus sylvestris*, *Picea sitchensis*, and *Picea abies*)	Short-rotation forests posed lower fire risks due to their susceptibility to the most damaging hazards. A close-to-nature system presented the lowest overall fire risk, while intensive even-aged systems posed the highest risk, regardless of species composition and bioclimatic factors.
9	Wright et al. (2012)	Dergholm, Southwest Victoria, Australia	*E. arenacea*	The edge effect significantly influenced photosynthesis, transpiration, and stomatal conductance, which were higher in edge trees compared to the interior due to greater light interception. Branch water potential was also significantly higher at the edges. While a previous study found edge effects to be 200% greater than in forest interiors, the current study reported only a 50% difference. The edge effect was less apparent during summer. Water availability likely contributed to higher transpiration at pasture edges than at plantation edges during the summer, resulting in significantly greater annual transpiration.
10	Feikema et al. (2010)	Not specified	*E. regnans*	Eucalyptus forest structure (i.e., stocking, LAI) and water balance components, including transpiration, evapotranspiration, rainfall interception, and runoff, were analysed through simulations spanning over 230 years for trends and magnitude. However, further validation is needed to predict how forest characteristics (age and structure) affect catchment water balances in relation to fire, land use change, and climate change.

(continued)

Table 3 (continued)

NO	Author	Location	Species	Main findings
11	Moreira et al. (2009)	Portugal	*Eucalyptus* spp.	Different land cover types exert different influences on fire events. Shrublands were identified as fire-prone land cover, unlike annual crops, permanent crops, and agroforestry systems. Eucalyptus land was less susceptible than conifer plantations, while broadleaved forests were the least fire-prone. Different climates, management, ignition patterns, firefighting strategies, and regional availability influence the susceptibility of land cover to fire.
12	De Faria et al. (2008)	Eastern Coastal Plain Region of Brazil	*Eucalyptus* spp.	The use of a subsoiler and planting hole did not affect the free light organic matter (FLOM) in Eucalyptus stands. However, the use of disc harrows reduced FLOM, resulting in lower concentrations of carbon (C), humic substances (HS), and total organic carbon (TOC) between planting rows. Stand age and less intensive soil preparation increased FLOM, though age did not affect the C content of HS or TOC.

P. patula; however, total litter input and the proportion of fine woody litter (branches and coarse roots) were significant factors (Lemma et al. 2007).

Edge effects have been identified as influential in plant physiological responses. Photosynthesis, transpiration, and stomatal conductance were higher in edge trees compared to interior trees, likely due to increased light interception and significantly higher branch water potential at the forest edge (Wright et al. 2012). On the other hand, the physiological performance of selected species should be a key consideration in tropical plantation development (Schönbeck et al. 2023; Kallarackal and Somen 2008). This is because transpiration rates vary with site characteristics (e.g., soil water availability) and species traits (e.g., rooting depth, LAI, and stomatal conductance). For instance, *E. grandis* demonstrates good stomatal control, with efficient transpiration and minimal reduction in photosynthetic rate even under dry season conditions (Kallarackal and Somen 1997a). Mielke et al. (2000) further identified that stomatal conductance is primarily controlled by photosynthetic active radiation (PAR), vapor pressure deficit (VPD), atmospheric CO_2 concentration, and leaf water potential. Net photosynthesis, meanwhile, is influenced by PAR and PVD.

In Brazil, water deficits affected the Eucalyptus plantation, causing growth gaps of 87% and 67% in the Southeast and North regions, respectively. In order to mitigate this issue, Elli et al. (2019) suggested the use of drought-tolerant genotypes, wider spacing with deeper subsoiling, and improved soil management practices. The authors also highlight the importance of monitoring forest fires, windstorms, and waterlogging as part of an integrated mitigation strategy.

Despite the positive contribution of Eucalyptus to microclimatic regulation and environmental functions, the large-scale conversion of natural areas into plantations, particularly with highly flammable exotic species such as Pinus and Eucalyptus, has been reported to pose fire threats to the surrounding ecosystem. For instance, Bowman et al. (2019) reported that 44% of *Pinus* and Eucalyptus plantations were burned in Chile, correlating with the highest fire severities in three decades. This highlights the vulnerability of these plantations to fire incidents, which are predicted to increase in frequency and severity.

However, contrasting findings have emerged. Alegria and Pedro (2019) reported that in Portugal, the expansion of Eucalyptus had no significant impact on increasing the burned area. They further noted that fire size was not primarily determined by forest composition and that the type of forest had only a minor influence on fuel hazard. In addition, short-rotation forests posed lower fire risks due to their susceptibility, while close-to-nature plantation types had a low fire risk. In contrast, the intensive even-aged system had the greatest fire risk, regardless of species composition or bioclimatic conditions (Jactel et al. 2012). This is further supported by Paelmo et al. (2016), a study from the first search string, who found that plantation management systems influence microclimate conditions, where short-rotation forestry-based agroforestry affected air and soil temperature and wind velocity. When the same system was adopted using *A. mangium* and *E deglupta*, it showed beneficial site impacts. These findings suggest that fire incidents are more closely related to management intensity and species homogeneity. Thus, there is a need to include heterogeneity as the basis of plantation development to reduce fire risks.

Different land cover types exert different influences on fire events, where shrublands are more fire-prone than annual crops, permanent crops, and agroforestry systems. Multiple other factors also determine the land cover susceptibility to fire, such as climate, management, ignition patterns, and firefighting strategies (Zacharakis and Tsihrintzis 2023; Moreira et al. 2009). In terms of plant selection, Moreira et al. (2009) reported that Eucalyptus plantations were less susceptible to fire than conifer plantations, while broadleaved forests were the least fire-prone of all.

Forest fire and land use and land cover change can affect the soil moisture and temperature, thereby influencing key soil processes (Arunrat et al. 2023; Arora and Chaudhry 2017). Changes in soil moisture and temperature directly affect microbial activity, leading to increased decomposition rate of soil organic matter and CO_2 efflux (Arora and Chaudhry 2017). Sound management practices are therefore essential for Eucalyptus plantations to mitigate soil fertility loss due to nutrient deficiency, organic matter depletion, or erosion (Alegria and Pedro 2019).

4 Conclusion

This review examined the literature related to *Eucalyptus* plantations, microclimate and forest fire studies. The extensive literature review revealed that diverse Eucalyptus species have been studied across various research topics. Findings indicate that Eucalyptus plantations offer functional microclimate mitigation; however, they also exhibit vulnerability to fire incidents. Therefore, extensive research on this vulnerability and fire-related risks associated with the species is crucial to re-evaluate their suitability as a key forest plantation species.

Moreover, the low number of returned articles in both searches suggests that further research is needed to investigate the impact of Eucalyptus trees on microclimate and forest fire events. Since microclimate significantly impacts ecosystem functioning, including the physiological performance, biodiversity, hydrology, soil processes, and forest fire incidents, deepening our understanding of these interactions is essential. Although microclimate research related to Eucalyptus has been conducted since 1974, limited research has focused on monitoring its environmental effects in detail. Importantly, establishing a clearer link between microclimate dynamics and forest fires is critical to support the development of mitigation strategies, especially in combating more intense meteorological hazards such as drought, extreme weather and fire occurrences. It is also recommended to expand the literature review by including multiple scholarly databases to capture a broader scope of existing research. Despite its limitation, this current review offers valuable insights for researchers, forest plantation stakeholders and practitioners, including local communities, to inform decision-making at various levels.

Acknowledgments This work was supported by the Trans-disciplinary Research Grant Scheme (TRGS) under the Ministry of Higher Education (MOHE), Malaysia (TRGS/1/2018/UPM/01/2/2 and TRGS/1/2018/UPM/01/2/4). This work is also part of the research carried out under the University-Driven Research Program (UDRP) titled "Sustainable Management and Utilization of Tropical Plant Resources."

References

Alegria C, Pedro N, Horta M d C, Roque N, Fernandez P (2019) Ecological envelope maps and stand production of eucalyptus plantations and naturally regenerated maritime pine stands in the central inland of Portugal. For Ecol Manag 432:327–344

Arnold JP, Fonseca CR (2011) Herbivory, pathogens, and epiphylls in Araucaria Forest and ecologically-managed tree monocultures. For Ecol Manag 262(6):1041–1046

Arora P, Chaudhry S (2017) Dependency of rate of soil respiration on soil parameters and climatic factors in different tree plantations at Kurukshetra, India. Trop Ecol 58:3

Arunrat N, Sereenonchai S, Kongsurakan P, Yuttitham M, Hatano R (2023) Variations of soil properties and soil surface loss after fire in rotational shifting cultivation in Northern Thailand. Front Environ Sci 11:1213181

Baqir M, Kothari R, Singh RP (2019) Fuel wood consumption, and its influence on forest biomass carbon stock and emission of carbon dioxide. A case study of Kahinaur, district Mau, Uttar Pradesh, India. Biofuels 10(1):145–154

Bowman DMJS, Moreira-Muñoz A, Kolden CA, Chávez RO, Muñoz AA, Salinas F, González-Reyes Á et al (2019) Human–environmental drivers and impacts of the globally extreme 2017 Chilean fires. Ambio 48(4):350–362

Buchanan G, Kennedy J (2009) Optimal selection of plantation tree species accounting for climate change. La Trobe University, Melbourne

Castagnolli L, Boggiani FS, Alberto J, de Lima M, Lima T, Tonello KC (2023) Hydrological properties of litter in different vegetation types: Implications for ecosystem functioning. Hydrology 10(8):165

Cordero Rivera A (2017) Large scale eucalypt plantations associated to increased fire risk. PeerJ Preprints 5:e3348v1

de Faria GE, de Barros NF, de Novais RF, da Silva IR, Neves JCL (2008) Organic carbon and fractions of soil organic matter at different distances from eucalyptus stem. Scientia Forestalis 36(80):265–277

do Nascimento MI, Poggiani F, Durigan G, Iemma AF, da Silva Filho DF (2010) The effectiveness of Eucalyptus barrier in containing the edge effect on a subtropical forest fragment in the state of São Paulo, Brazil. Scientia Forestalis 38(86):191–203

Elli EF, Sentelhas PC, de Freitas CH, Carneiro RL, Alvares CA (2019) Assessing the growth gaps of Eucalyptus plantations in Brazil–Magnitudes, causes and possible mitigation strategies. For Ecol Manag 451:117464

Evans J (2009) The multiple roles of planted forests. In: Planted forests: uses, impacts, and sustainability, pp 61–90

Feikema P, Morris J, Beverly C, Lane PNJ, Baker T (2010) Using 3PG+ to simulate longterm growth and transpiration in Eucalyptus regnans forests. In: International congress on environmental modelling and software modelling for environment's sake

Fernandes PM, Monteiro-Henriques T, Guiomar N, Loureiro C, Barros AMG (2016) Bottom-up variables govern large-fire size in Portugal. Ecosystems 19(8):1362–1375

Fernandes PM, Guiomar N, Rossa CG (2019) Analysing eucalypt expansion in Portugal as a fire-regime modifier. Sci Total Environ 666:79–88

Food and Agriculture Organization of the United Nations (2020) The state of the world's forests 2020, p 13

Fugère V, Kasangaki A, Chapman LJ (2016) Land use changes in an afrotropical biodiversity hotspot affect stream alpha and beta diversity. Ecosphere 7(6):e01355

Jactel H, Branco M, Duncker P, Gardiner B, Grodzki W, Langstrom B, Moreira F et al (2012) A multicriteria risk analysis to evaluate impacts of forest management alternatives on forest health in Europe. Ecol Soc 17(4):52

Kallarackal J, Somen CK (1997a) An ecophysiological evaluation of the suitability of Eucalyptus grandis for planting in the tropics. For Ecol Manag 95(1):53–61

Kallarackal J, Somen CK (1997b) Water use by Eucalyptus tereticornis stands of differing density in southern India. Tree Physiol 17(3):195–203

Kallarackal J, Somen CK (2008) Water loss from tree plantations in the tropics. Curr Sci 94:201–210

Korhonen J, Nepal P, Prestemon JP, Cubbage FW (2021) Projecting global and regional outlooks for planted forests under the shared socio-economic pathways. New For 52(2):197–216

Lai G, Zhang Y (2025) Increased atmospheric aridity and reduced precipitation drive the 2023 extreme wildfire season in Canada. Geophys Res Lett 52(6):e2024GL114492. https://doi.org/10.1029/2024GL114492

Lemma B, Kleja DB, Olsson M, Nilsson I (2007) Factors controlling soil organic carbon sequestration under exotic tree plantations: a case study using the CO2Fix model in southwestern Ethiopia. For Ecol Manag 252(1–3):124–131

Liu Z, Wu J, Zhou L, Lin Y, Fu S (2012) Tree girdling effect on bacterial substrate utilization pattern depending on stand age and soil microclimate in Eucalyptus plantations. Appl Soil Ecol 54:7–13

Lozano-García B, Muñoz-Rojas M, Parras-Alcántara L (2017) Climate and land use changes effects on soil organic carbon stocks in a Mediterranean semi-natural area. Sci Total Environ 579:1249–1259

Mann ML, Batllori E, Moritz MA, Waller EK, Berck P, Flint AL, Flint LE, Dolfi E (2016) Incorporating anthropogenic influences into fire probability models: Effects of human activity and climate change on fire activity in California. PLoS One 11(4):e0153589

Mielke MS, Oliva MA, de Barros NF, Penchel RM, Martinez CA, da Fonseca S, de Almeida AC (2000) Leaf gas exchange in a clonal eucalypt plantation as related to soil moisture, leaf water potential and microclimate variables. Trees 14(5):263–270

Moreira F, Vaz P, Catry F, Silva JS (2009) Regional variations in wildfire susceptibility of land-cover types in Portugal: implications for landscape management to minimize fire hazard. Int J Wildland Fire 18(5):563–574

Paelmo RF, Villancio VT, Paelmo GG (2016) Soil and microclimate amelioration of Short Rotation Forestry-based agroforestry in Cuenca, Batangas, Philippines. J Int Soc Southeast Asian Agric Sci 22(1):111–122

Payn T, Carnus J-M, Freer-Smith P, Kimberley M, Kollert W, Liu S, Orazio C, Rodriguez L, Silva LN, Wingfield MJ (2015) Changes in planted forests and future global implications. For Ecol Manag 352:57–67

Pereira DA, Noble A, Ian R (1989) Studies of water balance in eucalypt plantations in Portugal. Headwater Control IAHS Publication No. 174. International Association of Hydrological Sciences (IAHS), Wallingford, UK, pp. 23–31

Pfeifer M, Boyle MJW, Dunning S, Olivier PI (2019) Forest floor temperature and greenness link significantly to canopy attributes in South Africa's fragmented coastal forests. PeerJ 7:e6190

Schiller G, Karschon R (1974) Microclimate and recreational value of tree plantings in deserts. Landsc Plan 1(3):329–337

Schönbeck L, Arteaga M, Mirza H, Coleman M, Mitchell D, Huang X, Ortiz H, Santiago LS (2023) Plant physiological indicators for optimizing conservation outcomes. Conserv Physiol 11(1):coad073

Stam Å, Anttila J, Pellikka P, Rikkinen J (2020) Sensitivity of tropical pendant bryophytes: results from a translocation experiment along an elevation gradient. Ann Bot Fenn 57(1–3):71–81

Tjoelker M, Pfautsch S, Aspinwall M, Drake J, Choat B, Tissue D, Burykin T (2013) Putting the puzzle together: investigating hydraulic functioning and water transport at high spatial resolution in tall trees. In: IX international workshop on Sap Flow, vol 991, pp 245–251

Wan S, Liu Z, Chen Y, Zhao J, Ying Q, Liu J (2019) Effects of Lime Application and Understory Removal on Soil Microbial Communities in Subtropical Eucalyptus L'Hér. Plantations. *Forests* 10(4):338

Wang F, Zou B, Li H, Li Z (2014) The effect of understory removal on microclimate and soil properties in two subtropical lumber plantations. J For Res 19(1):238–243

Williams JJ, Newbold T (2020) Local climatic changes affect biodiversity responses to land use: a review. Divers Distrib 26(1):76–92

Wright AJ, Francia RM (2024) Plant traits, microclimate temperature and humidity: a research agenda for advancing nature-based solutions to a warming and drying climate. J Ecol 112(11):2462–2470

Wright TE, Tausz M, Kasel S, Volkova L, Merchant A, Bennett LT (2012) Edge type affects leaf-level water relations and estimated transpiration of *Eucalyptus arenacea*. Tree Physiol 32(3):280–293

Wyse SV, Etherington TR, Hulme PE (2022) Quantifying the risk of non-native conifer establishment across heterogeneous landscapes. J Appl Ecol 59(6):1608–1618

Zacharakis I, Tsihrintzis VA (2023) Integrated wildfire danger models and factors: a review. Sci Total Environ 899:165704

Zhao J, Wan S, Shenglei F, Wang X, Wang M, Liang C, Chen Y, Zhu X (2013) Effects of understory removal and nitrogen fertilization on soil microbial communities in Eucalyptus plantations. For Ecol Manag 310:80–86

Zhao J, Wan S, Zhang C, Liu Z, Zhou L, Fu S (2014) Contributions of understory and/or overstory vegetations to soil microbial PLFA and nematode diversities in eucalyptus monocultures. PLoS One 9(1):e85513

Zhou GY, Yan JH (2003) Hydrological impacts of reforestation with eucalyptus and indigenous species in southern China. Informatore Botanico Italiano 35(1):82–92

Zhou GY, Morris JD, Yan JH, Yu ZY, Peng SL (2002) Hydrological impacts of reafforestation with eucalypts and indigenous species: a case study in Southern China. For Ecol Manag 167(1–3):209–222

Tech-Driven Strategies for Agroforestry Soil Conservation: Integrating Smart Technologies and Eucalyptus-Based Applications

Darshini Rawichandran and Rosazlin Abdullah

Abstract Soil degradation, characterized by erosion, nutrient depletion, compaction, and climate variability, profoundly threatens sustainable agriculture and food security worldwide. Agroforestry systems, which synergistically integrate trees, crops, and livestock, offer a regenerative approach to soil conservation and ecosystem restoration. Among the tree species utilized, *Eucalyptus* has become one of the most widely adopted in agroforestry systems due to its rapid growth, ecological adaptability, and multifunctional uses ranging from timber and fuelwood to environmental services. This chapter advances the discourse by introducing Agroforestry 2.0, an innovation-driven paradigm that merges ecological knowledge with frontier technologies to strengthen soil conservation in *Eucalyptus*-based systems. Using tools such as geographic information systems (GIS), remote sensing, the Internet of Things (IoT), artificial intelligence (AI), digital twin simulations, and bioinformatics, these combined strategies allow for accurate assessments, flexible land-use planning, and up-to-date monitoring of soil health and fertility. Drawing from case studies across tropical and semiarid regions, this chapter demonstrates how technology-enhanced *Eucalyptus* agroforestry can mitigate degradation, increase microbial diversity, and optimize nutrient cycling within multifunctional landscapes. Furthermore, it explores the socioeconomic and institutional enablers needed for large-scale adoption among smallholders. Ultimately, this chapter envisions a future in which tech-enabled *Eucalyptus* agroforestry catalyzes climate resilience, ecosystem regeneration, and inclusive rural development.

D. Rawichandran · R. Abdullah (✉)
Institute of Biology, Faculty of Science, Universiti Malaya, Kuala Lumpur, Malaysia

Centre for Research in Biotechnology for Agriculture Research (CEBAR), Universiti Malaya, Kuala Lumpur, Malaysia
e-mail: rosazlin@um.edu.my

S. Mohd Razali et al. (eds.), *Sustainable Eucalyptus Forestry in Southeast Asia*, Forestry Sciences 88, https://doi.org/10.1007/978-3-032-21343-3_4

1 Introduction

Soil degradation poses a critical global challenge, threatening agricultural productivity, biodiversity, and ecosystem stability. For example, unsustainable practices such as deforestation, overgrazing, monocropping, and excessive chemical inputs have severely degraded over 33% of the world's soil (FAO 2020). These practices lead to soil erosion, nutrient depletion, and reduced water retention, impairing the ability of soils to support plant growth and ecosystem services. In particular, regions such as the tropics and semiarid areas are especially vulnerable to nutrient leaching, erosion, and water scarcity, exacerbating food insecurity and climate change vulnerability (Swift et al. 2004; Wall et al. 2004).

Agroforestry offers a dynamic and sustainable solution to these challenges by integrating trees, crops, and sometimes livestock within a single land-use system. Specifically, this approach stabilizes soils, reduces erosion, and enhances soil fertility through increased organic matter and nutrient cycling. Agroforestry systems mimic natural ecosystems and use alley cropping and silvopastoral systems to foster biodiversity, sequester carbon, and create resilient farming landscapes. Additionally, these systems play a big role in fighting climate change and making soil stronger by improving its physical features, such as how well it absorbs water and its overall structure, which are important for dealing with water shortages in at-risk areas (Carroll et al. 2004; Wardle et al. 2004).

The integration of modern technologies into agroforestry practices further enhances their potential. For example, tools such as geographic information systems (GISs), remote sensing, and wireless sensor networks enable real-time monitoring of soil and crop health, allowing precise resource management. Additionally, drones and satellite imagery provide large-scale insights into spatial variability, whereas Internet of Things (IoT) devices gather data to refine sustainable practices. By combining these advancements with traditional knowledge, agroforestry creates a precision-driven framework that enhances productivity and ecological balance (Carroll et al. 2004; Ramil Brick et al. 2022).

Ultimately, leveraging the synergy between ecological practices and cutting-edge technologies positions agroforestry as a transformative approach to soil conservation. This integration addresses the root causes of soil degradation, ensuring food security, climate resilience, and sustainable development. The potential of agroforestry to blend innovations with traditional systems presents a vital pathway for restoring soil health and protecting ecosystems for future generations (Zomer et al. 2009).

2 The Silent Erosion: Understanding the Multifaceted Challenges of Soil Degradation

Soil degradation is a pressing global issue with far-reaching impacts on agriculture, particularly in regions that experience heavy rainfall and steep slopes. One of the primary drivers of soil degradation is soil erosion, which is caused predominantly by wind and water. In tropical regions such as the Philippines and Indonesia, soil erosion has been worsened by deforestation and unsustainable agricultural practices, leading to the loss of vital nutrient-rich topsoil. According to Barrios (2007), such erosion significantly reduces soil fertility, which, in turn, diminishes crop yields. In areas where natural processes disrupt soil replenishment, maintaining agricultural productivity becomes even more challenging. Studies by the FAO (2020)), Swift et al. (2004), and Wardle et al. (2004) confirm that soil erosion is one of the leading factors in the decline of agricultural output in regions, particularly those where traditional farming methods fail to restore or preserve soil health.

In addition to erosion, nutrient depletion is another closely connected problem. Nutrients such as nitrogen, phosphorus, and potassium are vital for plant growth, but they are often leached into deeper soil layers or lost entirely via improper farming techniques. Unsustainable agricultural practices such as monocultures and overgrazing worsen nutrient loss, disrupting the natural nutrient cycle and increasing the difficulty of replenishing essential minerals. Barrios (2007) highlighted that nutrient depletion is particularly problematic in degraded lands, where the soil can no longer support healthy plant growth. For example, in sub-Saharan Africa, continuous cropping without proper replenishment of soil nutrients has led to poor soil quality, lower crop yields, and increased food insecurity. Studies by FAO (2020), Zomer et al. (2009), and Wardle et al. (2004) reported that nutrient depletion, when combined with soil erosion, further reduces soil fertility and impedes sustainable farming practices. Integrated soil management techniques that address both nutrient recovery and overall soil health are necessary; without them, nutrient depletion will continue to decrease productivity.

Another major challenge is soil compaction, particularly in farming systems where livestock are integrated with crop production. In silvopastoral systems, livestock trampling compacts the soil, reducing its porosity and preventing proper water infiltration and root growth. These deficiencies can lead to waterlogging and the development of anaerobic conditions, both of which harm plant and microbial life. According to Wardle et al. (2004), livestock movement in South American pasturelands has caused severe soil compaction, which reduces water drainage and hinders plant growth. The long-term impact of soil compaction reduces land productivity and degrades overall soil quality. Barrios (2007), FAO (2020), and Zomer et al. (2009) emphasized the importance of soil compaction in soil degradation, particularly in grazing systems where traditional farming practices often fail to mitigate livestock-induced damage to soil health. To prevent soil compaction and its negative effects on land productivity, sustainable agricultural systems that integrate trees, crops, and livestock, such as agroforestry, are essential.

Moreover, the impacts of climate change are increasingly complicating soil degradation. Extreme weather events such as droughts and floods are becoming more frequent and severe, worsening soil conditions. Droughts reduce soil moisture, weaken plant health, and cause the soil structure to deteriorate, whereas floods accelerate soil erosion and nutrient leaching, further depleting soil fertility. Swift et al. (2004) explained that these extreme weather events disrupt the natural balance of soil systems, causing long-term damage. In Southeast Asia, for example, climate change has led to unpredictable weather patterns, making it even more difficult for farmers to maintain soil health. Studies by Barrios (2007), FAO (2020), and Zomer et al. (2009) demonstrate that climate change exacerbates soil degradation by altering rainfall patterns, increasing the frequency of extreme weather events, and placing additional pressure on already vulnerable agricultural systems. As a result, traditional farming methods are no longer sufficient, and more resilient systems, such as agroforestry, are needed to adapt to these changing climate conditions.

While traditional soil conservation methods such as terracing, contour farming, and mulching have been used for centuries to combat soil erosion and improve soil fertility, these practices are often labor intensive and require ongoing maintenance, making them difficult to apply on a larger scale. According to the FAO (2020), although these techniques can be effective in localized settings, they are often impractical for larger farming landscapes with limited resources. For example, in parts of Africa, while contour farming has been used to prevent erosion of hilly terrain, the labor demands and continuous upkeep make it challenging for smallholder farmers to implement these methods sustainably. Furthermore, traditional methods often fail to address deeper issues such as nutrient depletion and soil microbial imbalances. As noted by Wardle et al. (2004) and Zomer et al. (2009), the solution lies in more integrated and innovative soil management practices, such as agroforestry, which combines trees, crops, and livestock to restore soil fertility in a more holistic way.

Another significant barrier to improving soil health is the lack of knowledge and training on modern soil conservation techniques. Many smallholder farmers continue to rely on outdated practices that do little to increase long-term soil fertility. Barrios (2007) highlighted that in regions such as Central America, farmers often underutilize advanced techniques such as biochar application, deep-rooted tree integration, and the use of microorganisms, all of which can significantly improve soil structure and nutrient retention. The absence of knowledge transfer, training, and technology perpetuates the cycle of soil degradation, as farmers remain unable to adopt more effective practices that could improve soil health and increase productivity. FAO (2020), Zomer et al. (2009), and Wardle et al. (2004) reported that without proper training and access to resources, smallholder farmers cannot make the necessary changes to improve soil fertility and sustain agricultural productivity in the long term.

In addition to knowledge gaps, socioeconomic factors such as financial constraints and insecure land tenure prevent the implementation of long-term soil management practices. Zomer et al. (2009) argue that in areas with limited financial resources or insecure land ownership, farmers are reluctant to invest in practices that require substantial upfront costs but deliver benefits only in the long term. This issue is particularly common in parts of Africa and Asia, where financial support and agricultural extension services are often lacking. Without secure land tenure and access to capital, farmers are unable to commit to practices such as agroforestry, even when these practices provide significant ecological and productivity benefits. These systemic constraints call attention to the need for more comprehensive and financially accessible solutions that integrate environmental and economic benefits.

Soil degradation, driven by factors such as erosion, nutrient depletion, compaction, and the effects of climate change, presents substantial challenges to sustainable agriculture. While traditional soil conservation methods have been valuable in specific contexts, they are often insufficient to address the broader scope of these issues. Agroforestry offers a more holistic solution by integrating trees, crops, and livestock to restore soil fertility, prevent erosion, and increase climate resilience. By overcoming the limitations of traditional methods and incorporating modern techniques, agroforestry can play a crucial role in combating soil degradation and ensuring the long-term sustainability of agricultural systems (Raskin and Osborn 2019).

3 The Agroforestry Renaissance: Tradition Redefined Through Innovation

Agroforestry is a dynamic and multifunctional land-use system that integrates trees, crops, and sometimes livestock into a single sustainable framework. By merging traditional agricultural practices with modern innovations, agroforestry provides comprehensive solutions to global challenges such as soil degradation, climate change, and food insecurity. It is an efficient approach that promotes sustainable farming practices and simultaneously maximizes ecological and economic benefits.

At the heart of agroforestry lies the principle of sustainability (Fig. 1), which emphasizes the long-term productivity of both agricultural and forestry systems while preserving environmental quality. Agroforestry achieves sustainability through practices that minimize soil erosion, conserve water, and reduce dependence on chemical fertilizers and pesticides. Sheldrick and Auclair (2000) and Smith et al. (2013) highlighted how agroforestry systems significantly reduce the environmental footprint of agriculture, making them central to global sustainable development efforts.

Biodiversity is another vital principle. Agroforestry systems integrate a wide variety of species, trees, shrubs, crops, and livestock, to create a more resilient agricultural system capable of withstanding environmental shocks such as extreme weather events and pest outbreaks. It also provides important ecosystem services, including pollination and natural pest control. Augère-Granier (2020) and Mosquera-Losada et al. (2018) reported that biodiversity within agroforestry systems fosters ecological balance and supports farmers' economic stability. Moreover, by optimizing the synergies between trees, crops, and livestock, agroforestry enhances resource

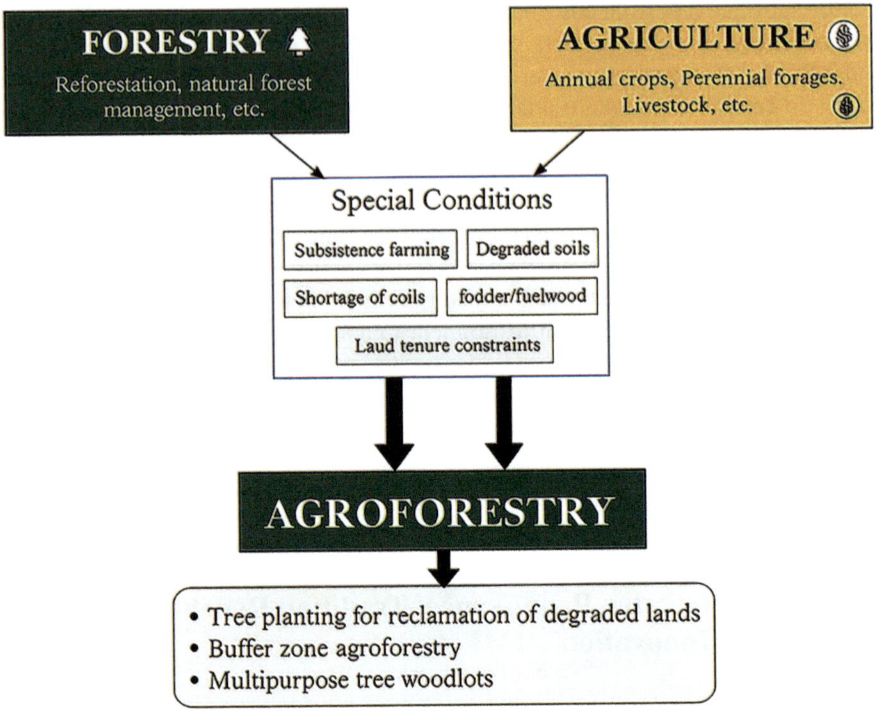

Fig. 1 Concept of agroforestry

sharing, improving overall productivity. For example, trees provide shade that reduces heat stress on livestock, whereas their roots improve soil structure and water infiltration, benefiting crops (Rao et al. 1997).

Agroforestry is inherently resource efficient, leveraging the complementary functions of its components to minimize resource waste and improve input utilization. These systems utilize sunlight, water, and soil nutrients, thereby contributing to both the economic viability of agricultural practices and the overarching goal of environmental sustainability (Bayala and Prieto 2020). Trees, in particular, contribute to this efficiency by sequestering carbon, regulating soil temperature, and improving water retention, which are vital for mitigating climate change and ensuring agricultural sustainability (Smith et al. 2013).

Agroforestry systems come in many forms, each offering distinct benefits depending on the local context (Fig. 2). Silvopastoral systems combine trees or shrubs with livestock grazing. These systems provide shelter for animals, improve soil health through manure recycling, and help reduce feed costs (Mosquera-Losada et al. 2018). Silvopastoral systems, or alley cropping, involve intercropping trees with annual or perennial crops. This practice enhances biodiversity and improves soil fertility through nutrient cycling (Perks et al. 2018). Forest farming systems leverage shaded environments, making them ideal for growing specialized crops

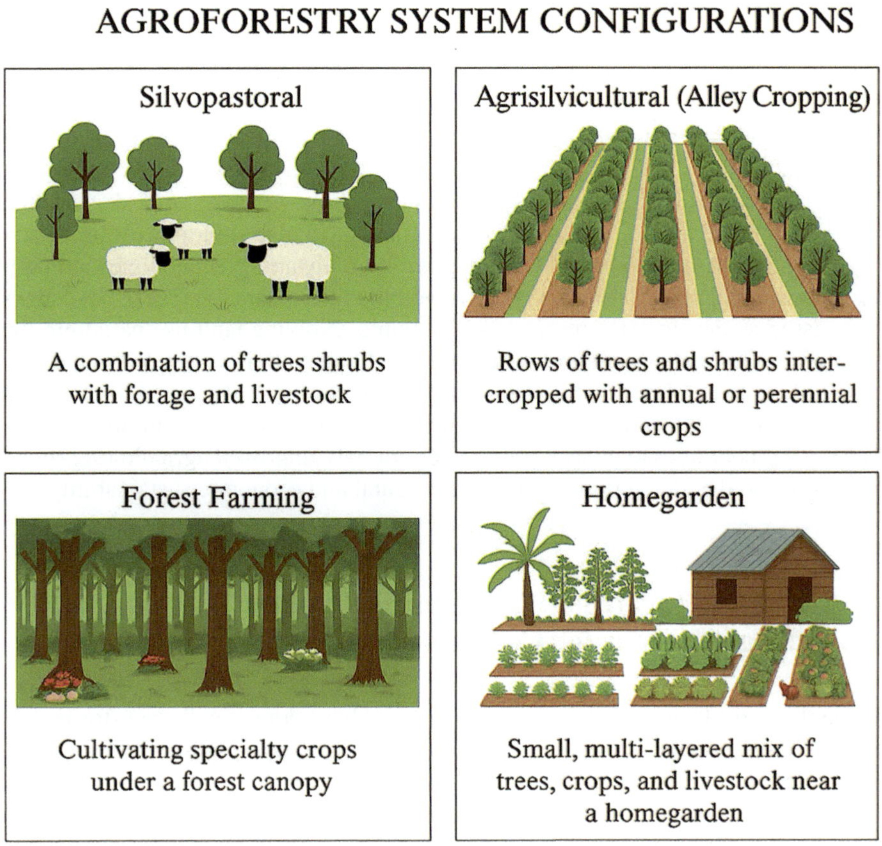

Fig. 2 Agroforestry system configurations. AI-generated image created with Google DeepMind, Gemini 2.5 Pro, 2025. © 2025 Darshini Rawichandran. Used with permission

such as mushrooms and medicinal plants. This approach conserves forest ecosystems and provides economic opportunities for farmers (Mosquera-Losada et al. 2018). Additionally, urban agroforestry solutions, such as home gardens, integrate trees, shrubs, and vegetables into small spaces, improving household food security and local microclimates (Sheldrick et al. 2000).

In response to growing pressures from population growth, urbanization, and environmental degradation, agroforestry has evolved into "Agroforestry 2.0." This modern iteration merges traditional ecological principles with cutting-edge technologies, overcoming long-standing limitations. The limitations include factors such as labor intensity and inefficiency. Important new ideas include using biochar, which helps make soil more fertile, balances pH levels, and increases the variety of microbes, tackling problems with keeping nutrients and maintaining healthy soil over time (Dong et al. 2024). Geospatial technologies such as GIS and remote sensing allow for detailed mapping of soil erosion risks and nutrient levels, enabling targeted conservation strategies (Bayala and Prieto 2020). Artificial intelligence (AI)-powered systems analyze large datasets to predict and mitigate risks, such as nutrient depletion and pest outbreaks, improving decision-making (Wolfert et al. 2017). Drones and IoT devices enhance efficiency by providing real-time monitoring of soil and crop health, ensuring precise resource management (Ramil Brick et al. 2022).

Agroforestry 2.0 thus addresses persistent issues such as soil degradation, resource inefficiency, and climate change, offering scalable solutions for food security and ecosystem restoration. The integration of advanced technologies strengthens the resilience of agroforestry systems, helping them adapt to modern challenges.

In essence, agroforestry represents a dynamic, evolving land-use paradigm that redefines traditional agricultural wisdom through innovation. Grounded in the principles of sustainability, biodiversity, integration, resource efficiency, and climate resilience, it offers a compelling path forward for regenerative agriculture. As agroforestry 2.0 continues to evolve, it promises not only improved agricultural productivity but also the global pursuit of environmental and economic sustainability.

4 Smart Soil Strategies: Bridging Technology and Agroforestry for Conservation

Technological advancements in soil management have dramatically transformed the way we conserve and enhance soil health in agroforestry systems. By integrating modern technologies, such as remote sensing, geographic information systems (GISs), the Internet of Things (IoT), and big data analytics, these innovations have played a critical role in ensuring the sustainability of agroforestry landscapes while increasing productivity. In agroforestry, where land use is multifunctional and requires tailored approaches, these technologies enable data-driven decisions that align ecological conservation with agricultural productivity (Sheldrick et al. 2000) (Table 1).

One essential tool in agroforestry is GIS, which helps provide detailed, high-resolution maps of soil conditions. These maps are vital for identifying areas prone to erosion, nutrient depletion, and compaction, enabling targeted interventions. For instance, GIS mapping has been used in hilly agroforestry areas to check for soil erosion risks and to carefully position plants that stop topsoil from washing away, which helps keep agroforestry systems sustainable (Ramil Brick et al. 2022). Remote sensing technologies also complement GIS by allowing for large-scale monitoring of soil properties, thus helping farmers make informed decisions about land use and vegetation management in agroforestry.

IoT-based soil sensors have become integral to precision agriculture in agroforestry systems. These sensors monitor crucial soil parameters such as moisture, pH, temperature, and nutrient levels in real time. The data gathered by these sensors support more efficient irrigation, fertilization, and crop management in agroforestry

Table 1 Technology in agroforestry soil conservation

Technology	Description	Application in agroforestry	Country	Researchers
Geographic information systems (GIS)	GIS provides detailed, high-resolution maps of soil conditions, identifying areas prone to erosion, nutrient depletion, and compaction.	We use GIS mapping to control soil erosion, manage nutrients, and strategically place vegetation.	Philippines, Thailand	Ramil Brick et al. (2022)
Remote sensing	Large-scale monitoring of soil properties enables farmers to make informed decisions on land use and vegetation management.	It is utilized to oversee vast agroforestry regions and aid in making well-informed choices for managing vegetation.	Brazil, India, Mexico	Keswani et al. (2019)
IoT-based soil sensors	The sensors are designed to monitor real-time soil parameters such as moisture, pH, temperature, and nutrient levels, enabling efficient irrigation and fertilization.	These sensors are utilized for effective water management and nutrient optimization in tree-crop systems.	India, Kenya, Australia	Sheldrick et al. (2000)
Big data analytics	Analyzes large datasets to uncover patterns and predict soil health issues like nutrient depletion or soil compaction.	It enhances agroforestry practices by forecasting factors such as the impact of drought on soil organic matter and microbial activity.	Kenya, Ethiopia, Brazil	Sishodia et al. (2020)
Artificial intelligence (AI)	Artificial intelligence (AI) uses predictive models to assess nutrient deficiencies in crops and optimize fertilizer application while minimizing environmental harm.	AI models predict nutrient levels and optimize the use of fertilizers for both crops and trees, reducing environmental impact.	China, Mexico, India	Lehmann et al. (2020)

(continued)

Table 1 (continued)

Technology	Description	Application in agroforestry	Country	Researchers
Digital twin technology	Creates virtual replicas of physical systems to simulate soil dynamics and evaluate the effects of agroforestry practices on soil health and productivity.	Simulates soil dynamics in agroforestry to guide land-use decisions and long-term sustainability.	Germany, Ethiopia, Chile	Sishodia et al. (2020)
Bioinformatics	Improves understanding of soil microbiomes and supports the introduction of drought-tolerant microbes to enhance soil resilience.	Enhance soil resilience by integrating drought-tolerant microbes into agroforestry systems to improve water retention and nutrient availability.	Japan, Kenya, India	Maruyama et al. (2024) and Awasthi et al. (2019)
Biochar and activated carbon	Improves soil fertility and structure by improving nutrient and water retention; biochar is used for soil restoration.	Help restore soil fertility in agroforestry by improving nutrient and water retention	Indonesia, Kenya	Lehmann et al. (2020) and Zhang et al. (2024)
Precision technologies	Automated pyrolysis for biochar production, GPS-guided biochar applications, and IoT sensors for composting streamline soil management.	Precision technologies enhance the efficiency of biochar production and composting, thereby enabling scalability and large-scale implementation in agroforestry.	Brazil, India, Kenya	Zomer et al. (2009)
Digital platforms	Mobile applications and online tools provide resources for best practices, training, and collaboration between farmers, researchers, and policymakers.	The digital platforms assist smallholder farmers in integrating traditional practices with modern technologies, thereby enhancing soil health and productivity.	Indonesia, Vietnam, Thailand	Barrios (2007)

systems. In India and Kenya, IoT-enabled systems have been used to optimize water and nutrient management, ensuring that irrigation and fertilization are adapted to meet the specific needs of tree-crop combinations within agroforestry systems (Barapatre and Patel 2019). Similarly, in Australia, IoT moisture sensors combined with automated irrigation systems have enabled farmers to manage water use efficiently, improving soil health and increasing yields across both crops and trees (Keswani et al. 2019).

Big data analytics and machine learning have further revolutionized soil management in agroforestry by uncovering patterns and predicting issues such as nutrient depletion or soil compaction. By looking at large amounts of data, predictive analytics uses information from sensors, satellite images, and weather records to predict things like how drought will affect soil health and tiny living things in the soil (Wolfert et al. 2017). These technologies are essential for optimizing agroforestry practices, including tree-crop combinations that increase nutrient cycling and water retention, ensuring overall sustainability in the system (Wall et al. 2004).

Artificial intelligence (AI) and machine learning have expanded their role in nutrient management within agroforestry. In countries such as China, AI models help forecast nutrient deficiencies in staple crops such as rice and wheat, optimizing fertilizer application while minimizing runoff due to the application of excessive fertilizer. In Mexico, AI platforms predict nutrient levels in agroforestry systems, helping to apply the right amount of fertilizer for both trees and companion crops. Additionally, in India, AI-based tools help assess tree growth and evaluate the impact of irrigation practices, improving water efficiency and minimizing pesticide use (Lehmann et al. 2020).

Digital twin technology, which creates virtual replicas of physical systems, is increasingly being used in agroforestry to simulate soil dynamics and evaluate the effects of various practices on soil health and crop productivity. In Germany, digital twins simulate different agroforestry practices to guide better land-use decisions, whereas in Ethiopia and Chile, such models have been instrumental in predicting the risks of soil erosion and assessing soil fertility trajectories, giving researchers invaluable knowledge about long-term sustainability (Bargués Tobella et al. 2014; Sishodia et al. 2020).

Bioinformatics has also emerged as a transformative tool for soil microbiome analysis, offering insights into microbial populations that underpin soil fertility. In Japan, bioinformatics tools are utilized to profile and analyze microbial communities, enabling more efficient management to support nutrient cycling (Maruyama et al. 2024). In Kenya and India, bioinformatics facilitates the introduction of drought-tolerant microbes, increasing water retention and nutrient availability in agroforestry systems under adverse climatic conditions (Awasthi et al. 2019; Maruyama et al. 2024).

Modern technologies are harnessed to improve traditional soil management practices such as biochar production, composting, cover cropping, and mulching. Biochar, which improves soil fertility by increasing nutrient and water retention, has been widely integrated into agroforestry systems. In Indonesia, it is used to restore nutrient-depleted soils (Lehmann et al. 2020), whereas in Kenya, it is applied to degraded soils, reducing the need for chemical fertilizers and promoting long-term soil health (Zhang et al. 2024). Activated carbon, a refined form of biochar, stabilizes the soil structure, prevents nutrient leaching, and contributes to soil sustainability in agroforestry systems (Sishodia et al. 2020).

Precision technologies such as automated pyrolysis for making biochar, GPS-guided biochar applications, and IoT sensors for composting help simplify soil management in agroforestry systems. These advancements make biochar production, composting, and cover cropping more efficient, scalable, and adaptable to large agroforestry operations (FAO 2020; Zomer et al. 2009). GPS-guided seeders and automated mulching machines are making it easier to use cover cropping and mulching on large farms in agroforestry systems.

Digital platforms, such as mobile applications and online tools, are becoming invaluable resources for agroforestry practitioners. These platforms provide access to best practices, training, and a collaborative space for farmers, researchers, and policymakers. In Southeast Asia, these digital platforms have supported smallholder farmers by integrating traditional agroforestry methods with modern technologies, leading to improved soil health and productivity outcomes (Barrios 2007). These platforms aggregate data from various technologies, such as IoT sensors, allowing real-time monitoring of soil moisture, temperature, and fertility, which are vital for ensuring optimal soil conservation and crop yields (Fig. 3) (Keswani et al. 2019).

In conclusion, bridging technology and agroforestry offers smart soil strategies that contribute to soil conservation, enhance agricultural productivity, and improve resilience to climate change. By integrating modern technologies with traditional agroforestry practices, these strategies ensure long-term sustainability, increasing productivity while maintaining ecological balance. These innovations preserve soil

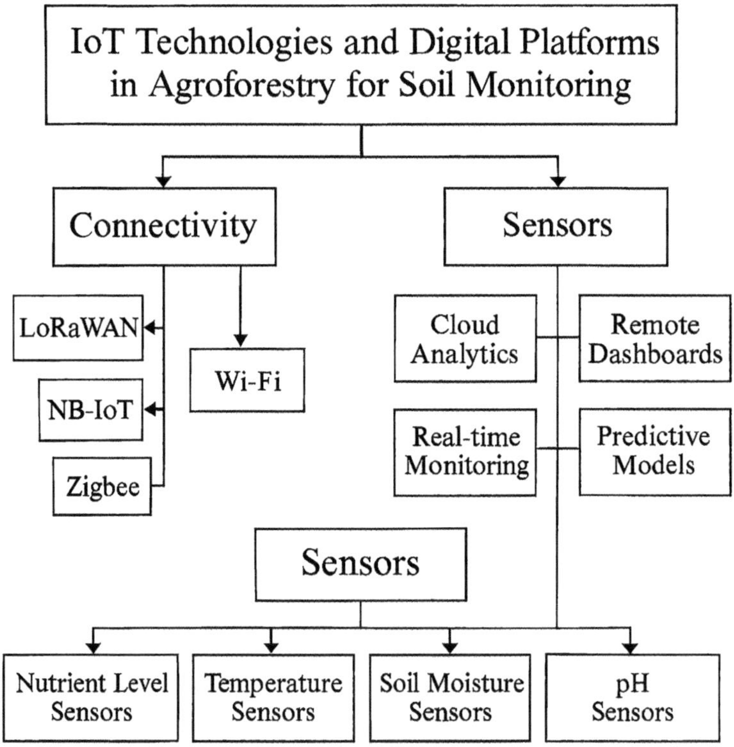

Fig. 3 IoT technologies and digital platforms used in agroforestry for soil monitoring

health and support agroforestry systems in meeting the demands of global food production and climate resilience.

5 Soil Health Dynamics and Sustainability in *Eucalyptus*-Based Agroforestry Systems

Eucalyptus species have emerged as some of the most extensively cultivated trees in agroforestry systems worldwide due to their rapid growth rate, adaptability across diverse ecological zones, and wide-ranging uses, from timber and fuelwood to ecological services. Their increasing adoption is primarily motivated by both economic and environmental incentives. Economically, they provide construction materials and generate substantial income opportunities for farmers. Ecologically, they contribute to carbon sequestration and soil improvement. The genus *Eucalyptus* encompasses species with distinct root and canopy architectures, which influence belowground processes and surface microclimates differently. Consequently, understanding how *Eucalyptus* affects soil properties is essential, particularly as its expansion into farmlands raises questions about long-term sustainability.

Research indicates that *Eucalyptus*-based agroforestry systems can enhance soil properties, including pH, cation exchange capacity (CEC), and organic carbon under specific conditions. However, these systems also present challenges, including nutrient depletion and allelopathic effects, which may emerge under poor management (Daka and Geleta 2023). Studies have shown that intercrops grown alongside *Eucalyptus* clones experience varying degrees of yield reduction, maize, for instance, often demonstrates significant growth decline under *Eucalyptus* canopies (Ramesh et al. 2023). Despite these drawbacks, *Eucalyptus* plays a crucial socioeconomic role by supplying fuelwood and reducing pressure on natural forests, particularly in regions like Rwanda, where its cultivation supports energy security and rural livelihoods (Ndayambaje and Mohren 2011).

5.1 Soil Physio-chemical Properties Under Eucalyptus *Agroforestry*

Eucalyptus plantations influence several soil physical characteristics that directly affect soil health and productivity. Research from the Gidami District in West Ethiopia revealed that soil bulk density and exchangeable acidity under *Eucalyptus* did not significantly differ from those in control plots, indicating that *Eucalyptus* does not inherently compact soil (Daka and Geleta 2023). Moreover, increases in soil moisture and pH near tree canopies have been observed, primarily due to litter deposition and shading effects that lower evapotranspiration rates. Similarly,

Eucalyptus saligna plantations in the eastern Democratic Republic of Congo were found to enhance the proportion of large macroaggregates, improving soil structure, fertility, and erosion resistance through enhanced pore connectivity (Kataka et al. 2024).

Nevertheless, findings remain context-dependent. In certain central Ethiopian regions, soils under *Eucalyptus* woodlots displayed lower organic carbon and moisture content compared to other land uses, suggesting potential vulnerability to degradation under poor management (Haile et al. 2022). Thus, while *Eucalyptus* can positively influence soil physical properties, careful site-specific assessments are crucial to balance its benefits with potential risks to soil structure.

Chemically, *Eucalyptus* agroforestry impacts soil pH, nutrient dynamics, and organic matter differently across sites and depths. For instance, in western Oromia, Ethiopia, soil pH and CEC were higher near tree bases, likely due to litter decomposition and root exudation (Daka and Geleta 2023). Yet, other studies report opposite trends, with lower pH under dense *Eucalyptus* stands, reflecting the complexity of soil–tree interactions. Advancements in soil biogeochemistry research have also emphasized the importance of monitoring nitrogen cycling and greenhouse gas emissions in managed ecosystems. Butterbach-Bahl et al. (2016) demonstrated that soil nitrogen transformations play a crucial role in regulating nitrous oxide emissions and nutrient losses in agricultural and forest soils. Their work underscores the importance of improved nutrient management strategies to minimize environmental impacts and enhance soil sustainability. Integrating smart technologies such as soil nutrient sensors and digital monitoring tools can facilitate real-time monitoring of soil nitrogen dynamics and optimize fertilizer application in agroforestry systems. Nitrogen availability often decreases because of tree uptake competition, whereas potassium and organic matter contents tend to rise near root zones. In semi-arid Telangana, India, *Eucalyptus* plantations enhanced surface soil fertility through increases in organic carbon and phosphorus, demonstrating positive correlations between CEC and macronutrient availability (Varshitha et al. 2023). Furthermore, *Eucalyptus* litter decomposition promotes phosphate-solubilizing microbes and mycorrhizal fungi, facilitating phosphorus cycling Corra et al. 2015).

In terms of soil organic matter (SOM) and carbon sequestration, *Eucalyptus* systems contribute substantial organic inputs through leaf litter and root biomass. Although carbon content under *Eucalyptus* woodlots is typically lower than under native forests, it remains higher than that of continuously cropped lands, indicating a relative carbon sequestration advantage (Daka and Geleta 2023). Moreover, integrated *Eucalyptus–coffee* systems have been shown to store significant carbon across multiple pools, aboveground, belowground, litter, and soil, thereby contributing to climate change mitigation (Toru and Kibret 2019). Such findings underscore the potential of *Eucalyptus*-based agroforestry to enhance soil carbon stocks and support environmental sustainability when appropriately managed (Kumar et al. 2025).

5.2 Eucalyptus *Influence on Soil Fertility Dynamics*

In addition to nutrient cycling, root–soil interactions play an important role in regulating soil chemistry and microbial processes. Fujii and Hayakawa (2021) reported that plant roots release organic compounds that influence soil acidity and nutrient availability in forest ecosystems. These root exudates interact with soil microorganisms and contribute to complex rhizosphere processes that affect nutrient mobilization and soil fertility. Understanding these interactions can help improve soil conservation strategies in agroforestry systems through the integration of plant–microbe interactions and sustainable land management practices. For example, litter decomposition and microbial transformations beneath *Eucalyptus* trees play a crucial role in driving nutrient cycling and maintaining soil fertility in tropical ecosystems. As *Eucalyptus* leaves decompose, they release mineralized nutrients such as nitrogen (N) and phosphorus (P), which become available to both trees and associated intercrops. Microbial communities, particularly arbuscular and ectomycorrhizal fungi and phosphate-solubilizing microorganisms, contribute significantly to nutrient mobilization and uptake, sustaining fertility in otherwise nutrient-limited tropical soils (Corra et al. 2015). However, *Eucalyptus* species' efficient nutrient uptake often leads to localized nitrogen depletion, which can negatively affect adjacent crops (Daka and Geleta 2023).

Recent findings by Islam et al. (2025) demonstrated that *Eucalyptus camaldulensis* boundary plantations significantly influence soil nutrient dynamics in rice fields in Bogra District, Bangladesh. The study reported that soils collected closer to *Eucalyptus* trees (0 m) exhibited reduced concentrations of N, P, K, S, Cu, and Zn, while concentrations gradually increased with distance from the tree line. These patterns were attributed to the trees' extensive and deep root systems, which compete for soil nutrients and moisture with nearby crops. The authors also observed higher organic matter and electrical conductivity near *Eucalyptus* boundaries due to continuous litter deposition and root exudation, whereas soil moisture levels declined because of high evapotranspiration rates and water uptake efficiency (Islam et al. 2025).

Soil microbial processes under *Eucalyptus* plantations exhibit both beneficial and inhibitory characteristics. Enhanced root exudation and litter inputs stimulate enzymatic activities associated with nitrogen, phosphorus, and sulfur cycling (Gianfreda 2015). Nonetheless, allelopathic compounds such as phenolics and terpenoids released from *Eucalyptus* roots and decomposing litter can suppress certain microbial taxa, potentially reducing microbial diversity and nutrient turnover (Daka and Geleta 2023). The study by Islam et al. (2025) further revealed the accumulation of trace metals, such as Ni, As, Cd, and Pb in soils near *Eucalyptus* boundaries, suggesting biochemical interactions between root exudates and heavy-metal availability that may alter microbial functioning and nutrient cycling.

From an agroecological perspective, *Eucalyptus* plantations exert a dual influence on soil fertility and crop productivity. While they contribute to enhanced soil structure, organic matter accumulation, and microbial stimulation, they also monopolize water and nutrient resources, often reducing yields of intercrops such as maize and sorghum, especially under dense canopy conditions (Ramesh et al. 2023). Strategic management practices such as wider tree spacing (3–5 m), incorporation of leguminous intercrops, and integrated nutrient management can mitigate these adverse effects and promote complementary resource utilization within *Eucalyptus*-based systems (Kaur et al. 2001; Islam et al. 2025).

Overall, *Eucalyptus*-based agroforestry systems exhibit both regenerative and competitive traits. They enhance soil biological activity through litter decomposition and nutrient recycling, but simultaneously intensify competition for soil nutrients and water. Balanced management strategies that integrate intercrops, microbial inoculants, and optimized spacing are therefore vital for sustaining soil health, microbial diversity, and long-term productivity in *Eucalyptus*-influenced agroecosystems.

5.3 Tech-Driven Monitoring of Soil Properties in **Eucalyptus** *Agroforestry*

Modern technological tools are transforming how soil conditions under *Eucalyptus* agroforestry are studied and managed. Remote sensing and GIS technologies allow researchers to assess spatial variations in soil moisture, organic matter, and erosion susceptibility using multispectral and hyperspectral imagery. These approaches enable the identification of soil quality gradients and facilitate targeted soil conservation planning (Pazhanivelan et al. 2025). Similarly, GIS-based mapping supports continuous monitoring of infiltration, degradation, and recovery processes in *Eucalyptus* systems (Kataka et al. 2024).

Advancements in soil sampling and laboratory analyses, such as GPS-enabled sample collection and high-precision analytical instruments, enhance data accuracy and reproducibility (Varshitha et al. 2023). Coupled with data analytics software, these tools allow the development of predictive models to interpret large datasets, thereby strengthening the understanding of soil dynamics under agroforestry (Alexandre et al. 2015). Furthermore, the integration of Internet of Things (IoT) sensors facilitates real-time monitoring of soil parameters like pH, moisture, and nutrient concentrations, enabling data-driven precision management (Bhanuwanti et al. 2024). Such smart systems not only optimize irrigation and fertilization but also provide alerts for drought or leaching events, supporting resilient and efficient *Eucalyptus*-based agroforestry (Kumari et al. 2024; Regmi and Paudel 2024).

5.4 *Soil Conservation Techniques in* Eucalyptus*-Based Agroforestry*

Erosion control and soil conservation are fundamental to sustainable *Eucalyptus* cultivation. Techniques such as terracing, contour planting, and stone bund installation have proven highly effective in reducing surface runoff and enhancing water infiltration, particularly in hilly areas of Ethiopia and Kenya (Birhan and Tekalign 2022; Mutio et al. 2021). Complementary vegetative measures, such as cover crops and grass strips, help stabilize soil, maintain ground cover, and enrich organic matter inputs, with studies in Brazil showing up to a 50% reduction in soil loss under such integrated systems (Sone et al. 2019).

In addition, biopore technology has emerged as an innovative approach to improve infiltration and aeration. In Indonesia, biopores created within *Eucalyptus* plantations enhanced water retention and reduced flooding, while the use of organic amendments like compost and manure further improved soil structure and fertility (Erif et al. 2025; Birhan and Tekalign 2022). Collectively, these conservation practices not only sustain soil productivity but also strengthen the ecological resilience of *Eucalyptus* agroforestry systems (Kumar et al. 2025).

Moreover, diversified agroforestry models, such as alley cropping and agrosilvopastoral systems, have been developed to harmonize soil conservation with income generation (Yusuf and Astiko 2025). These integrated designs optimize spatial and temporal resource use while minimizing competition between trees, crops, and livestock. Adoption of such models in East Africa has reached over 80% in some regions, driven by clear socioeconomic and ecological benefits (Korir et al. 2022; Altieri et al. 2011).

5.5 *Economic and Environmental Implications of* Eucalyptus *Agroforestry on Soil*

Economically, the productivity of *Eucalyptus* agroforestry is strongly linked to its impact on soil fertility and crop yields. While intercrops like maize often experience yield declines of up to 40% under *Eucalyptus* (Ramesh et al. 2023), system-level profitability can still improve when appropriate crop combinations are selected. The inclusion of complementary species increases land-use efficiency and stabilizes income through diversified outputs (Kaur et al. 2001). Beyond direct income, *Eucalyptus* agroforestry enhances fuelwood supply, employment, and food security, particularly in regions with limited forest resources (Korir et al. 2022).

From an environmental perspective, *Eucalyptus* contributes substantially to soil carbon storage through litter deposition and root turnover, enhancing soil structure and fertility while mitigating climate change (Daka and Geleta 2023). These systems also foster biodiversity by creating habitats for soil fauna, mycorrhizal fungi, and beneficial arthropods (Udawatta et al. 2019). Nevertheless, replacing natural forests with *Eucalyptus* monocultures may alter species composition and ecosystem functionality, demanding careful planning to maintain ecological balance (Malmer et al. 2009).

Challenges persist in maintaining a balance between soil and water. The high water and nutrient uptake rates of *Eucalyptus* can lead to soil nutrient depletion and reduced groundwater availability in semi-arid regions (Hadgu et al. 2009). Similarly, allelopathic compounds may suppress understory vegetation and disrupt soil microbial functions (Haile et al. 2022). These constraints highlight the need for integrated nutrient management and species diversification to sustain productivity and ecological integrity.

5.6 Integrating Eucalyptus *Agroforestry with Climate-Smart Agriculture Practices*

Eucalyptus agroforestry aligns seamlessly with climate-smart agriculture principles by enhancing carbon sequestration, moderating microclimates, and improving soil resilience. Integrating trees with crops reduces greenhouse gas emissions compared to monocropping and contributes to national emission reduction targets (Kumar et al. 2025). Consequently, many countries, including India, have incorporated agroforestry into climate mitigation policies to expand forest cover and improve carbon stocks (Basu 2014).

Moreover, *Eucalyptus*-based systems enhance soil water retention and regulate microclimate, making them particularly valuable in dryland regions prone to erratic rainfall (Yusuf and Astiko 2025). The use of digital tools such as remote sensing and IoT sensors provides adaptive solutions for monitoring soil moisture and erosion risks in real time (Regmi and Paudel 2024; Pazhanivelan et al. 2025).

Beyond environmental resilience, *Eucalyptus* agroforestry promotes socioeconomic stability by diversifying livelihoods, improving food security, and empowering rural communities. Farmers in Kenya and Ethiopia often prefer *Eucalyptus* for its fast returns and compatibility with boundary planting, balancing economic and ecological objectives (Korir et al. 2022). Scaling up these practices, however, requires enhanced access to finance, capacity-building programs, and participatory knowledge-sharing mechanisms (Altieri et al. 2011).

6 Next-Generation Approaches: Shaping the Future of Tech-Enhanced Agroforestry for Soil Conservation

Technological innovations are transforming the future of agroforestry soil conservation, offering promising solutions to enhance soil health, fertility, and sustainability. Advances in artificial intelligence (AI) and machine learning (ML) are revolutionizing soil management by allowing farmers to predict and optimize nutrient cycles, water usage, and pest control more accurately (Lehmann et al. 2020). These technologies reduce reliance on chemical fertilizers by providing data-driven recommendations that improve soil fertility while minimizing environmental impact (Ennaji et al. 2023). The adoption of Internet of Things (IoT) sensors in agroforestry systems is also expanding rapidly, allowing real-time monitoring of soil moisture, nutrient levels, and overall health. This enables farmers to make informed decisions regarding irrigation and fertilization, improving resource use efficiency and reducing waste (Keswani et al. 2019). Additionally, digital twin technology has emerged as a powerful tool, enabling the creation of virtual models of soil dynamics and agroforestry performance. By simulating various management practices, digital twins help identify the most effective strategies for enhancing soil fertility, preventing erosion, and increasing biodiversity without the need for physical field trials (Bargués Tobella et al. 2014; Sishodia et al. 2020). Bioinformatics further strengthens these advancements by improving understanding of the soil microbiome's role in nutrient cycling and soil vitality, leading to the development of microbial inoculants that improve soil fertility and resilience (Maruyama et al. 2024).

Despite their immense potential, challenges remain in adopting these technologies widely. High implementation costs, particularly for smallholder farmers, continue to limit accessibility, especially in developing regions (Zomer et al. 2009). Financial constraints, insecure land tenure, and limited technical literacy also discourage long-term investment in soil conservation technologies (Ennaji et al. 2023). To address these barriers, targeted financial support, digital infrastructure development, and comprehensive capacity-building initiatives are needed to assist farmers in adopting and utilizing these innovations effectively (Lehmann et al. 2020). Collaboration between technology developers, agricultural scientists, and local communities ensures that innovations are contextually relevant, affordable, and adapted to local environmental conditions.

Within *Eucalyptus*-based agroforestry systems, next-generation technologies hold significant potential to address long-standing soil-related challenges. However, there remain critical research gaps in understanding the long-term and site-specific effects of *Eucalyptus* on soil nutrient dynamics and ecosystem health. Studies reveal limited understanding of species-specific allelopathic mechanisms and their temporal and spatial variability, as well as incomplete knowledge of interactions with diverse soil microbial and faunal communities across different environments (Daka and Geleta 2023). These uncertainties restrict the ability to optimize *Eucalyptus* management for soil conservation and ecosystem balance. Regional variability in soil responses and the scarcity of longitudinal studies further constrain the generalization of findings, calling for integrative and multidisciplinary approaches that combine ecological, agronomic, and socioeconomic perspectives (Haile et al. 2022; Kataka et al. 2024).

Emerging technologies are addressing these limitations by enabling advanced soil monitoring and management in *Eucalyptus* agroforestry. Unmanned aerial vehicles (drones) equipped with multispectral sensors can capture fine-scale spatial variation in soil and vegetation, complementing ground-based measurements. IoT sensor networks enable continuous monitoring of soil moisture, temperature, and nutrient levels, supporting real-time adaptive management practices (Pazhanivelan et al. 2025). Similarly, advances in molecular biology and bioinformatics have enhanced the profiling of soil microbial communities and their functional capacities, enabling the correlation of microbial diversity indices with soil fertility and organic matter content (Gianfreda 2015; Maruyama et al. 2024). These tools, when integrated with AI and digital twin modeling, enable precision agroforestry management that enhances soil quality, reduces degradation, and improves climate resilience. However, technical challenges persist in managing large datasets, ensuring sensor affordability, and developing user-friendly analytical tools to guide farmers' decision-making (Kumari et al. 2024).

Policy and institutional support play a pivotal role in scaling up these technology-driven soil conservation practices. Effective implementation requires strong policy frameworks, subsidies, and extension programs to promote the adoption of sustainable land management tools. Capacity-building initiatives that train farmers and extension officers enhance the uptake of advanced technologies and support adaptive soil management (Korir et al. 2022). Collaboration across government agencies, research institutions, and private sectors fosters innovation dissemination, while participatory approaches that incorporate farmers' feedback ensure that technologies are socially acceptable and contextually appropriate (Altieri et al. 2011). Ultimately, a balanced strategy combining technological innovation, institutional support, and grassroots participation is crucial for the widespread implementation of sustainable and resilient agroforestry systems (Sittadewi et al. 2025).

With the right combination of technology, policy, and education, *Eucalyptus*-based agroforestry can become a model for climate-smart land management. The integration of AI, IoT, digital twins, drones, and bioinformatics provides a comprehensive framework for improving soil health, biodiversity, and productivity while mitigating the impacts of climate change. These next-generation technologies will enable the creation of intelligent, self-regulating agroecosystems that optimize soil conditions, enhance nutrient use efficiency, and sustain ecological balance, ultimately transforming agroforestry into a regenerative, adaptive, and future-ready system.

7 Conclusion

The integration of advanced technologies has profoundly transformed agroforestry soil conservation, marking a shift from conventional management to intelligent, data-driven practices that prioritize sustainability and resilience. Technologies such as artificial intelligence (AI), machine learning (ML), the Internet of Things (IoT), and digital twins now allow real-time monitoring of soil health, empowering farmers and researchers to make precise and timely decisions regarding nutrient management, irrigation, and pest control (Lehmann et al. 2020; Ennaji et al. 2023). These innovations go beyond traditional observation methods by enabling predictive modeling and proactive mitigation of soil degradation, thereby strengthening the resilience and productivity of agroforestry ecosystems. The introduction of smart biochar, microbial technologies, and blockchain-based systems further enhances soil fertility, boosts carbon sequestration, and ensures transparency and traceability in sustainable soil management practices. Together, these advancements not only address existing soil conservation challenges but also lay a robust foundation for scalable, long-term solutions that align with global sustainability goals (Sida 2018).

Future research and practice in agroforestry soil conservation must focus on developing soil-specific technological applications tailored to diverse agroecological zones. Innovations in bioinformatics, nanotechnology, and AI-driven models should be adapted to accommodate variations in soil type, structure, and climatic conditions to maximize efficiency and ecological compatibility (Maruyama et al. 2024). It is equally crucial that these technologies remain affordable and user-friendly, particularly for smallholder farmers who represent the backbone of agroforestry production in developing regions. Inclusive access and cost-effective implementation will ensure that digital transformation in soil conservation benefits all stakeholders equitably (Zomer et al. 2009). Additionally, the integration of renewable energy systems, such as solar-powered IoT sensors and smart irrigation, can further align soil conservation with sustainable energy transitions, reducing carbon footprints while improving resource efficiency. Community-centric and participatory platforms that engage farmers in co-designing, testing, and sharing technological knowledge are essential for fostering local ownership and enhancing adoption rates (Altieri et al. 2011).

In the context of *Eucalyptus*-based agroforestry, these technological advancements play a crucial role in optimizing soil conservation outcomes. *Eucalyptus* species exert complex and multifaceted influences on soil physico-chemical properties and fertility. Positive outcomes include increased soil moisture, higher pH, improved cation exchange capacity (CEC), and enhanced organic matter content, largely attributed to litter deposition and root–microbial interactions that improve soil aggregation and carbon sequestration (Daka and Geleta 2023). However, potential challenges such as nutrient depletion, allelopathic inhibition, and water competition necessitate careful management to prevent ecological imbalance (Varshitha et al. 2023). The magnitude of *Eucalyptus* impacts varies significantly based on site conditions, species selection, and management intensity, underlining the importance of localized and adaptive approaches that integrate ecological and economic considerations for sustained soil and livelihood benefits (Kumar et al. 2025).

To maximize soil conservation benefits, the implementation of integrated, technology-driven strategies in *Eucalyptus* agroforestry is essential. Remote sensing and Geographic Information Systems (GIS) allow comprehensive spatial mapping of soil properties and help identify areas requiring targeted intervention. IoT-based soil sensors and precision agriculture tools support adaptive nutrient and water management by providing continuous, real-time data (Pazhanivelan et al. 2025). Advanced laboratory analyses of soil samples, combined with field-level conservation practices such as biopore technology, organic amendments, and terracing, synergize to enhance soil structure, nutrient retention, and moisture balance

(Bhanuwanti et al. 2024). Tailoring these approaches to local agroecological and socioeconomic contexts through participatory and capacity-building frameworks is vital for their effective implementation (Yusuf and Astiko 2025).

Looking ahead, sustained progress in *Eucalyptus* agroforestry requires strong policy frameworks, long-term field research, and cross-disciplinary collaboration. Future research priorities should focus on understanding nutrient dynamics, allelopathic effects, and microbial interactions across different soil and climatic conditions, while technological innovation should aim to enhance affordability and usability of soil-monitoring tools (Sittadewi et al. 2025). Policymakers should strengthen institutional coordination, establish financial incentives, and develop open-access knowledge-sharing platforms that encourage adoption of sustainable land management technologies (Altieri et al. 2011). Integrating ecological science with social and economic research will provide a holistic foundation for sustainable intensification of *Eucalyptus* agroforestry (Regmi and Paudel 2024).

In conclusion, the future of agroforestry soil conservation lies at the intersection of technology, innovation, and inclusivity. The integration of AI, IoT, digital twins, and bioinformatics with renewable energy and participatory governance models will revolutionize soil management, enabling intelligent, resilient, and sustainable agroecosystems. Through collaborative research, farmer empowerment, and supportive policies, agroforestry systems, particularly those involving *Eucalyptus*, can become models of climate-smart, regenerative land management that safeguard soil health, enhance productivity, and promote environmental stewardship for generations to come.

Author Declaration on AI-Generated Content
Artificial intelligence (AI) tools were used solely for figure creation (specifically Fig. 2) using *Gemini 2.5 Pro* (Google DeepMind, 2025). The AI was not used for writing, data analysis, or interpretation. The author conceptualized the figure content and ensured full compliance with Springer Nature's AI usage and copyright policies.

Acknowledgments The author expresses heartfelt appreciation to the Institute of Biological Sciences, Faculty of Science, and the Centre for Research in Biotechnology for Agriculture (CEBAR), Universiti Malaya, for the invaluable mentorship, scholarly guidance, and encouragement that have greatly enriched the preparation of this book chapter. The collective expertise and unwavering support provided by colleagues and collaborators have been instrumental in shaping the intellectual depth and scientific quality of this work. The author also extends gratitude to Universiti Malaya for providing access to research facilities, digital resources, and an intellectually stimulating environment that made this scholarly contribution possible.

References

Alexandre MESEDO, Ednaldo CANR, Vitor CEADMB, Adilson PA, Alex DSB (2015) Evaluation and comparison of soil under integrated croplivestock-forest system in the southeast of gois, Brazil. Acad J. https://doi.org/10.5897/ajar2015.9735

Altieri MA, Funes-Monzote FR, Petersen P (2011) Agroecologically efficient agricultural systems for smallholder farmers: contributions to food sovereignty. Agron. Sustain. Dev 32:1–13. https://doi.org/10.1007/s13593-011-0065-6

Augère-Granier M-L (2020) Agroforestry in the European Union. European Parliamentary Research Service, Brussels. https://www.europarl.europa.eu/thinktank/en/document/EPRS_BRI(2020)651991

Awasthi MK, Sarsaiya S, Wainaina S, Rajendran K, Kumar S, Quan W, Duan Y, Awasthi SK, Chen H, Pandey A, Zhang Z, Jain A, Taherzadeh MJ (2019) A critical review of organic manure biorefinery models toward sustainable circular bioeconomy: technological challenges, advancements, innovations, and future perspectives. Renew Sust Energ Rev 111:115–131. https://doi.org/10.1016/j.rser.2019.05.017

Barapatre P, Patel JN (2019) The determination of soil moisture via various sensors for irrigation water management. Int J Innov Technol Explor Eng 8:576–582

Bargués Tobella A, Reese H, Almaw A et al (2014) The effect of trees on preferential flow and soil infiltrability in an agroforestry parkland in semiarid Burkina Faso. Water Resour Res 50:3342–3354. https://doi.org/10.1002/2013WR015197

Barrios E (2007) Soil biota, ecosystem services and land productivity. Ecol Econ 64(2):269–285. https://doi.org/10.1016/j.ecolecon.2007.03.004

Basu JP (2014) Agroforestry, climate change mitigation and livelihood security in India. N.Z. J. For. Sci. 44(Suppl 1):S11. https://doi.org/10.1186/1179-5395-44-s1-s11

Bayala J, Prieto I (2020) Water acquisition, sharing and redistribution by roots: applications to agroforestry systems. Plant Soil 453:17–28. https://doi.org/10.1007/s11104-019-04173-z

Bhanuwanti, Dar KA, Singh L, Saad A, Rai U, Tanwar H, Khatoon A (2024) Climate smart agriculture: innovating sustainable practices for a changing climate. J Sci Res Rep 30(11):568–576. https://doi.org/10.9734/jsrr/2024/v30i112585

Birhan T, Tekalign W (2022) Sustainable agricultural management of land using technology for soil and water conservation within the central rift valley, central Ethiopia. Appl Environ Soil Sci. https://doi.org/10.1155/2022/7329580

Butterbach-Bahl K, Sander BO, Pelster D, Díaz-Pinés E (2016) Quantifying greenhouse gas emissions from managed and natural soils. In: Rosenstock TS, Rufino MC, Butterbach-Bahl K et al (eds) Methods for measuring greenhouse gas balances and evaluating mitigation options in smallholder agriculture. Springer, Cham, pp 71–96. https://doi.org/10.1007/978-3-319-29794-1_4

Carroll ZL, Bird SB, Emmett BA, Reynolds B, Sinclair FL (2004) Can tree shelterbelts on agricultural land reduce flood risk? Soil Use Manag 20(3):357–359. https://doi.org/10.1111/j.1475-2743.2004.tb00381.x

Corra EJA, Carvalhais LC, Utida MK, Oliveira C, Scotti M (2015) Effect of plant species on p cycle-related microorganisms associated with litter decomposition and p soil availability: implications for agroforestry management. iForest 9:294–302. https://doi.org/10.3832/ifor1459-008

Daka YS, Geleta ZT (2023) Effect of eucalyptus tree on selected soil physico-chemical properties in gidami district, west Ethiopia. Research Square. (United States). https://doi.org/10.21203/rs.3.rs-3264978/v1

Dong X, Zhang P, Zang Y, Cao Y, Wang B, Ouyang L, Zhang S, Duan H, Zhang W, Li Y, Yan H, Gao Y, Chen Z, Zhang X, Li W, Li J, Wang W, Chen K, He C, Zhang X, Dai J, Qiao Y, Lin D, Wang J (2024) InternLM-XComposer2-4KHD: a pioneering large vision-language model handling resolutions from 336 pixels to 4K HD. Adv Neural Inf Proces Syst 37:42566–42592

Ennaji O, Vergütz L, El Allali A (2023) Machine learning in nutrient management: a review. Artif Intell Agric 9:1–11. https://doi.org/10.1016/j.aiia.2023.06.001

Erif LOM, Gandri L, Sudia L, Indriyani L, Kahirun K, Siwi LO, Teke J, Tuwu ER, Midi L, Hidayat H, Fitriani V (2025) Agroforestri: inovasi hijau dalam upaya mitigasi perubahan iklim di desa andalambe kecamatan tongauna utara kabupaten konawe. AKSILAR Akselerasi Luaran Pengabdian Masyarakat 2(2):87–94. https://doi.org/10.19184/aksilar.v2i2.5430

Food and Agriculture Organization of the United Nations (2020) The impact of soil change on ecosystem services. In: FAO (ed) Status of the world's soil resources (SWSR)—Main report. FAO, Rome, pp 179–210

Fujii K, Hayakawa C (2021) Root exudation and biodegradation of organic acids in a tropical forest soil under dipterocarp and pioneer trees. Plant Soil 469:213–226

Gianfreda L (2015) Enzymes of importance to rhizosphere processes. J. Soil Sci. Plant Nutr 15(2). https://doi.org/10.4067/s0718-95162015005000022

Hadgu K, Kooistra L, Rossing W, Bruggen AV (2009) Assessing the effect of faidherbia albida based land use systems on barley yield at field and regional scale in the highlands of tigray, Northern Ethiopia. Food Sec 1:337–350. https://doi.org/10.1007/s12571-009-0030-2

Haile G, Itanna F, Teklu B, Agegnehu G (2022) Variation in soil properties under different land use types managed by smallholder farmers in central Ethiopia. Taylor & Francis. https://doi.org/10.1080/27658511.2022.2093058

Islam MS, Ahmed S, Miah MMU, Mahiddin NA, Al Bakky A, Ismail Z, Sahlabji T, Idris AM (2025) Assessing the impact of eucalyptus trees on soil chemical properties in rice fields. Sci Rep 15(16262). https://doi.org/10.1038/s41598-025-01220-x

Kataka JM, Onwonga RN, Karanja N, Kamau S (2024) Tree species and macrofauna influence on soil aggregates and carbon in eastern Democratic Republic of Congo. Autonomous University of Yucatn. https://doi.org/10.56369/tsaes.5330

Kaur N, Singh B, Gill RIS (2001) Productivity and profitability of intercrops under four tree species throughout their rotation in north-western India. Indian J Agron 62(2):160–169. https://doi.org/10.59797/ija.v62i2.4295

Keswani C, Prakash OM, Bharti N, Vílchez JI, Sansinenea E, Lally RD, Borriss R, Singh SP, Gupta VK, Fraceto LF, de Lima R, Singh HB (2019) Readdressing the biosafety issues of plant growth promoting rhizobacteria. Sci Total Environ 690:841–852. Google Sites. https://doi.org/10.1016/j.scitotenv.2019.07.046

Korir KE, Sirmah P, Matonyei TK, Nampushi JSO (2022) Classification and socio-economic benefits of agroforestry systems in soin ward, kericho county, Kenya. East Afr J For Agrofor 5(1):252–268. https://doi.org/10.37284/eajfa.5.1.904

Kumar NA, Jattan P, Digambar WD, Mohapatra R, Yadav KK, Sharma C, Chaudhary S (2025) A review of soil organic carbon dynamics under regenerative agricultural practices. Int J Plant Soil Sci 37(9):142–153. https://doi.org/10.9734/ijpss/2025/v37i95693

Kumari N, Pandey AK, Singh A, Singh A (2024) A review of present advancement in climate smart agriculture techniques: adaptation for sustaining crop production under climate change. Arch Curr Res Int 24(7):52–63. https://doi.org/10.9734/acri/2024/v24i7828

Lehmann LM, Smith J, Westaway S et al (2020) Productivity and economic evaluation of agroforestry systems for sustainable production of food and nonfood products. Sustainability 12(13):5429. https://doi.org/10.3390/su12135429

Malmer A, Murdiyarso D, Bruijnzeel LA, Ilstedt U (2009) Carbon sequestration in tropical forests and water: a critical look at the basis for commonly used generalizations. Glob. Change Biol 16(2):599–604. https://doi.org/10.1111/j.1365-2486.2009.01984.x

Maruyama R, Yasumoto K, Mizusawa N, Iijima M, Yasumoto-Hirose M, Iguchi A, Hermawan OR, Hosono T, Takada R, Song K-H, Shinjo R, Watabe S, Yasumoto J (2024) Metagenomic analysis of the microbial communities and associated network of nitrogen metabolism genes in the Ryukyu limestone aquifer. Sci Rep 14:4356. https://doi.org/10.1038/s41598-024-54614-8

Mosquera-Losada MR, Santiago-Freijanes JJ, Rois-Díaz M, Moreno G, den Herder M, Palma JHN, Freese D, Rigueiro-Rodríguez A (2018) Agroforestry in Europe: a land management policy tool to combat climate change. Land Use Policy 78:603–613. https://doi.org/10.1016/j.landusepol.2018.06.052

Mutio J, Njoroge R, Kebeney S, Ngetich W, Churu H, Alkamoi B, Wamalwa F (2021) Dynamics of terrace adoption, propagation, benefits & challenges among dryland agro-pastoralist communities of west pokot. Authorea. https://doi.org/10.22541/au.164050501.16730417/v1

Ndayambaje JD, Mohren G (2011) Fuelwood demand and supply in Rwanda and the role of agroforestry. Agrofor Syst 83:303–320. https://doi.org/10.1007/s10457-011-9391-6

Pazhanivelan S, Lad SU, Selvakumar S, Ravikumar V, Ramesh AV, Karale OS, Jadhav RJ (2025) Assessment of climate change on soil erosion using geospatial techniques: a review. Int J Environ Climate Chang 15(6):191–209. https://doi.org/10.9734/ijecc/2025/v15i64883

Perks M, Khomik M, Bathgate S et al (2018) Agroforestry in Scotland-potential benefits in a changing climate. CXC Report

Ramesh KR, Deshmukh H, Sivakumar K, Guleria V, Umedsinh RD, Krishnakumar N, Thangamalar A, Suganya K, Kiruba M, Selvan T, Balasubramanian PBP, Ushamalini C, Thiyagarajan G, Vincent S, Rajeswari P, Bavish S, Riaz A, Kuppusamy S (2023) Influence of eucalyptus agroforestry on crop yields, soil properties, and system economics in southern regions of India. Sustainability 15(4):3797. https://doi.org/10.3390/su15043797

Ramil Brick ES, Holland J, Anagnostou DE et al (2022) A review of agroforestry, precision agriculture, and precision livestock farming, the case for a data-driven agroforestry strategy. Front Sens 3:998928. https://doi.org/10.3389/fsens.2022.998928

Rao SC, Rainer G, Miller EK (1997) Integration of what and where in the primate prefrontal cortex. Science 276(5313):821–824. https://doi.org/10.1126/science.276.5313.821

Raskin B, Osborn S (2019) The agroforestry handbook: agroforestry for the UK. Soil Association, Bristol

Regmi S, Paudel B (2024) Climate-smart agriculture: a review of sustainability, resilience, and food security. Arch Agric Environ Sci 9(4):832–839. https://doi.org/10.26832/24566632.2024.0904028

Sheldrick RD, Auclair D (2000) Origins of agroforestry and recent history in the UK. In: Hislop AM, Claridge JN (eds) Agroforestry in the UK, Forestry commission bulletin 122. Forestry Commission, Edinburgh, pp 7–16

Sheldrick R, Auclair D, Hislop M, Claridge J (2000) Origins of agroforestry and recent history in the UK. In: Hislop AM, Claridge J (eds) Agroforestry in the UK, Forestry Commission Bulletin 122. Forestry Commission, Edinburgh, pp 7–16

Sida TS (2018) Sustainable intensification of smallholder farming systems in Ethiopia: what roles can scattered trees play? Wageningen University. https://doi.org/10.18174/425918

Sishodia RP, Ray RL, Singh SK (2020) Applications of remote sensing in precision agriculture: a review. Remote Sens 12(19):3136. https://doi.org/10.3390/rs12193136

Sittadewi EH, Tejakusuma I, Handayani T, Santoso AD, Tohari A, Mulyono A, Zakaria Z, Budiman EB, Fatahillah HEH, Fitriani R (2025) Novel rap-landslide method for assessing agroforestry sustainability in landslide-prone areas. Resources 14(6):93. https://doi.org/10.3390/resources14060093

Smith J, Pearce BD, Wolfe MS (2013) Reconciling productivity with protection of the environment: is temperate agroforestry the answer? Renewa Agric Food Syst 28(1):80–92. https://doi.org/10.1017/S1742170511000585

Sone JS, Oliveira PTSD, Zamboni P, Vieira NO, Carvalho GA, Macedo MCM, Arajo ARD, Montagner DB, Sobrinho TA (2019) Effects of long-term crop-livestock-forestry systems on soil erosion and water infiltration in a brazilian cerrado site. Sustainability 11(19):5339. https://doi.org/10.3390/su11195339

Swift MJ, Izac AM, van Noordwijk M (2004) Biodiversity and ecosystem services in agricultural landscapes, are we asking the right questions? Agric Ecosyst Environ 104(1):113–134. https://doi.org/10.1016/j.agee.2004.01.013

Toru T, Kibret K (2019) Carbon stock under major land use/land cover types of hades sub-watershed, eastern Ethiopia. Carbon Balance Manag 14:7. https://doi.org/10.1186/s13021-019-0122-z

Udawatta RP, Rankoth LM, Jose S (2019) Agroforestry and biodiversity. Sustainability 11(10):2879. https://doi.org/10.3390/su11102879

Varshitha M, Chaitanya T, Neelima TL, Jayasree G (2023) Soil characteristics under different agroforestry land use systems in semiarid region. Int. J. Environ. Clim. Chang 13(9):562–573. https://doi.org/10.9734/ijecc/2023/v13i92271

Wall TD, Michie J, Patterson M, Wood SJ, Sheehan M, Clegg CW, West M (2004) On the validity of subjective measures of company performance. Pers Psychol 57(1):95–118. https://doi.org/10.1111/j.1744-6570.2004.tb02485.x

Wardle DA, Bardgett RD, Klironomos JN, Setälä H, van der Putten WH, Wall DH (2004) Ecological linkages between aboveground and belowground biota. Science 304(5677):1629–1633. https://doi.org/10.1126/science.1094875

Wolfert S, Ge L, Verdouw C, Bogaardt MJ (2017) Big data in smart farming– a review. Agric Syst 153:69–80. https://doi.org/10.1016/j.agsy.2017.01.023

Yusuf M, Astiko W (2025) Agroforestry as an ecological approach to support the sustainability of dryland agroecosystems in pringgabaya district, east lombok. J Biol Trop 25(4):5309–5324. https://doi.org/10.29303/jbt.v25i4.10016

Zhang S-H, Shen Y, Lin L-F, Tang S-L, Liu C-X, Fang X-H, Guo Z-P, Wang Y-Y, Zhu Y-C (2024) Effects of bamboo biochar on soil physicochemical properties and microbial diversity in tea gardens. PeerJ 12:e18642. https://doi.org/10.7717/peerj.18642

Zomer RJ, Trabucco A, Coe R (2009) Trees on farm: analysis of global extent and geographical patterns of agroforestry. ICRAF Working Paper no. 89. World Agroforestry Centre, Nairobi

Forest Management Perspectives

Challenges and Opportunities in Tropical Mangrove Plantations: Ecological Interactions with *Eucalyptus*

Mohammad Basyuni, Alfian Mubaraq, Andi Aznan Aznawi, Tien Wahyuni, and Tadashi Kajita

Abstract Mangroves are important coastal ecosystems with ecological, economic, and social value, including carbon sequestration, shoreline protection, and biodiversity. Indonesia has the world's largest mangrove area, but mangrove deforestation and degradation are driven by aquaculture, agriculture, and urban expansion. Indonesia has focused on smart plantations and the restoration of mangroves, and the challenges and opportunities of these efforts are reported here. The lack of knowledge among the general public, funding restrictions, conflicts of interest among stakeholders, and failure to consider technical aspects prevent the success of restoration efforts. Indonesia has pledged to rehabilitate 600,000 ha of mangroves through community-based programs backed by new technologies. There are opportunities to integrate mangrove restoration with carbon-offset markets and ecotourism initiatives, creating sustainable livelihoods for coastal communities while contributing to climate change resilience. Policies, such as the establishment of the Indonesian Mangrove Peat Restoration Agency (BRGM), show that large-scale implementation is possible but requires coordination among multiple stakeholders with robust policy frameworks and innovative financing mechanisms. Indonesia could be a world leader in sustainable mangrove management, address

M. Basyuni (✉)
Center of Excellence for Mangroves, Universitas Sumatera Utara, Medan, Indonesia

Department of Forestry, Faculty of Forestry, Universitas Sumatera Utara, Medan, Indonesia
e-mail: m.basyuni@usu.ac.id

A. Mubaraq · A. A. Aznawi
Center of Excellence for Mangroves, Universitas Sumatera Utara, Medan, Indonesia

Graduate School of Agricultural Science, Faculty of Agriculture, Universitas Sumatera Utara, Medan, Indonesia

T. Wahyuni
Research Center for Behavioral and Circular Economics, National Research and Innovation Agency (BRIN), Jakarta, Indonesia

T. Kajita
Iriomote Station, Tropical Biosphere Research Center, University of the Ryukyus, Taketomi, Okinawa, Japan

S. Mohd Razali et al. (eds.), *Sustainable Eucalyptus Forestry in Southeast Asia*, Forestry Sciences 88, https://doi.org/10.1007/978-3-032-21343-3_5

climate imperatives, and improve coastal communities. This chapter briefly discusses mangrove forests and *Eucalyptus* plantations to highlight their ecological interactions.

1 Introduction

Mangroves are critical ecosystems that provide numerous ecological, social, and economic benefits. Indonesia, home to the most extensive mangrove areas globally, plays a vital role in the conservation of these ecosystems. Mangroves are natural coastal protectors against erosion and storm surges and serve as important carbon sinks, supporting global climate change mitigation efforts (Donato et al. 2011; Alongi 2012). Despite these benefits, mangrove ecosystems in Indonesia face numerous challenges, particularly from anthropogenic pressures, such as land conversion for aquaculture, agriculture, and urban development, which have resulted in considerable deforestation and degradation (Giri et al. 2011).

Despite their importance, mangrove forests in Indonesia face immense challenges. Between 1980 and 2020, Indonesia lost nearly 40% of its mangrove cover due to aquaculture expansion, logging, and land reclamation for agriculture and urbanization (Spalding and Leal 2021). These losses are both ecological and economic, with coastal communities disproportionately affected by declining fisheries and increased vulnerability to natural disasters. The loss of mangroves affects the biodiversity and ecosystem services they provide, and the livelihoods of local communities that depend on them for sustenance and income.

Ecological challenges in mangrove plantations arise from improper restoration. Many efforts have relied on monoculture plantations, such as *Rhizophora* spp., which often fail in the long term because of their limited adaptability to site-specific conditions. For example, studies have shown that the survival rate of mangroves in degraded areas can reach 10–20% when ecological and hydrological factors, such as tidal inundation and sediment supply, are ignored (Lewis 2005; Primavera and Esteban 2008). Additionally, the loss of species diversity can compromise ecosystem resilience and reduce the capacity of plantations to provide habitat and carbon sequestration services (Friess et al. 2019).

Social challenges further complicate mangrove restoration. More than 70% of Indonesia's mangroves are located in areas with high poverty rates, where local communities depend on mangroves for timber, charcoal, and traditional aquaculture (Basyuni et al. 2022a). Conflicts between conservation objectives and community livelihoods often arise because mangrove conservation may limit the resources available to these communities, thereby limiting the effectiveness of plantation projects. Balancing the needs of communities with mangrove conservation is a complex and ongoing challenge in restoration efforts.

Despite these challenges, substantial opportunities exist in optimizing mangrove plantation efforts in Indonesia. Remote sensing and geospatial technologies have revolutionized mangrove monitoring, providing hope in adverse situations. For example, satellite data combined with machine-learning algorithms can detect mangrove degradation with up to 90% accuracy, allowing for targeted restoration interventions (Asbridge et al. 2016; Fatoyinbo et al. 2018). Integrating these tools with local ecological knowledge has effectively increased plantation success rates. Policy support also provides a strong foundation for scaling up mangrove plantations. The Indonesian government's National Mangrove Ecosystem Management Strategy (2021–2030) targets the restoration of 600,000 ha of mangroves by 2024, with a budget allocation exceeding IDR 3 trillion (approximately USD 200 million) (Ministry of Environment and Forestry 2021). International partnerships, such as the Blue Carbon Initiative, support these efforts, aligning mangrove conservation with carbon market opportunities.

Community-based approaches, such as the integration of mangrove plantations with ecotourism and sustainable aquaculture, have shown promise. These approaches show not only the potential of local communities in mangrove restoration but also the power of collective action. For example, in northern Sumatra, community-led mangrove restoration projects have resulted in a 25% increase in fishery yields while enhancing coastal protection (Basyuni et al. 2022a).

Silvicultural management is a prevalent method in mangrove restoration (Kusmana 2014; Kodikara et al. 2017) and involves the extensive planting of monocultured mangrove seedlings. This strategy frequently overlooks assessments of ecosystem habitat functionality and adaptive management. Evaluating the progress and success of restoration is crucial for identifying optimal and repeatable strategies for stakeholders and for understanding the underlying processes that facilitate the return of ecosystem function (Ellison 2008).

Restoration initiatives globally seek to rehabilitate coastal mangroves and improve their essential biodiversity and ecosystem functions. In Southeast Asia, mangrove restoration efforts have targeted abandoned aquaculture ponds and former palm oil plantations that supplanted mangroves. Currently, the criteria for restoration effectiveness are limited to seedling survival rates and plant community composition, emphasizing the need to identify other indicators to evaluate the functionality of restored ecosystems.

This review aims to critically synthesize current knowledge on the ecological, technical, and socio-economic challenges and opportunities associated with tropical mangrove plantations in Indonesia, with a particular emphasis on the ecological interactions between mangrove ecosystems and adjacent Eucalyptus plantations. Specifically, it seeks to (1) identify key factors influencing the success and sustainability of mangrove restoration and rehabilitation, (2) highlight existing knowledge gaps and policy limitations, and (3) propose integrated management strategies that balance ecological integrity, climate change mitigation, and community livelihoods. By addressing these objectives, the review provides a comprehensive foundation for enhancing sustainable mangrove management and informing future policy and research directions in tropical regions.

2 Challenges of Tropical Mangrove Plantations in Indonesia

2.1 *Degradation and Loss of Mangrove Habitats*

Mangrove deforestation and degradation in Indonesia, which has the world's largest amount of mangrove coverage, are occurring at alarming rates, primarily due to conversion to aquaculture, agriculture, and other land usage (Arifanti 2020; Cahyaningsih et al. 2022). Anthropogenic activities, such as mining, logging, and pollution, have led to substantial mangrove loss, changes in species composition, and ecosystem degradation (Cahyaningsih et al. 2022). Between 2009 and 2019, Indonesia lost an estimated 182,091 ha of mangroves, potentially releasing 182.6 Mton of CO_2 into the atmosphere (Arifanti et al. 2021). These losses contribute to climate change, increase vulnerability to disasters, and reduce community income (Utami et al. 2024). Management strategies that consider social, economic, and environmental factors, including increased community participation, sustainable spatial planning, law enforcement, and improved institutional frameworks, are necessary to address these issues (Utami et al. 2024; Arifanti 2020). Conservation and restoration efforts are crucial for mitigating climate change and preserving the biodiversity of these ecosystems (Arifanti et al. 2021).

2.2 *Extensive Deforestation for Aquaculture, Agriculture, and Urban Development Has Resulted in the Degradation of Mangrove Forests*

The primary drivers of extensive deforestation include conversion to aquaculture, agriculture, urban development, and palm oil plantations (Richards and Friess 2016; Basyuni and Sulistiyono 2018; Arifanti et al. 2021; Cahyaningsih et al. 2022) (Table 1). Between 2009 and 2019, Indonesia lost an estimated 182,091 ha of mangroves, resulting in potential emissions of 182.6 Mton CO_2e (Arifanti et al. 2021). In northern Sumatra, primary mangrove forests decreased by 61.21% between 1990 and 2015, and no reforestation was observed (Basyuni and Sulistiyono 2018), as depicted in Table 1. Across Southeast Asia, mangrove deforestation occurred at an average rate of 0.18% per year from 2000 to 2012, with aquaculture accounting for 30% of the total forest change (Richards and Friess 2016). These activities have reduced mangrove areas, caused changes in species composition and pollution, and threatened biodiversity and ecosystem services (Cahyaningsih et al. 2022).

Mangrove ecosystems are crucial for climate change mitigation through carbon storage and sequestration and provide resilience against climate-induced events (Arifanti 2020). Indonesia has implemented various management strategies to address this issue, including rehabilitation and community-based management, but faces challenges in terms of policy implementation and coordination among government agencies (Quevedo et al. 2023; Arifanti 2020).

Table 1 Drivers of mangrove loss in Indonesia

Drivers	Sites in Indonesia	Sources of references
Aquaculture expansion	Northern Java, North Sulawesi, and East Kalimantan	Bosma et al. (2012); Richards and Friess (2016); Basyuni and Sulistiyono (2018)
Urban development	Jakarta Bay, Bali, and Surabaya	Hakim et al. (2017)
Agricultural conversion	Sumatra, Kalimantan, and Papua	Friess et al. (2019); Eddy et al. (2021); Giri et al. (2022)
Illegal logging	West Kalimantan and North Sumatra	Obidzinski (2003); Basyuni et al. (2025)
Infrastructure development	DKI Jakarta	Slamet et al. (2020)
Palm oil plantations expansion	North Sumatra, Central Kalimantan, and West Papua	Hamilton and Friess (2016); Basyuni and Sulistiyono (2018)
Mining activities	East Kalimantan and Bangka-Belitung	Ardianto et al. (2020)
Sea-level rise and coastal erosion	Northern Java and Banda Aceh	Meilianda et al. (2019); Sagala et al. (2024)

However, mangrove rehabilitation efforts in Indonesia have yielded inconsistent results. While some programs have been successful, with increased mangrove forest areas observed in the Subang and Indramayu districts (Nursamsi and Komala 2017), approximately 79% of mangrove planting efforts across Indonesia have been unsuccessful because of low survival rates and inadequate land preparation (Suyadi et al. 2023).

Successful rehabilitation depends on various factors, including soil fertility, hydrological conditions, and species selection (Blake et al. 2021; Maiti et al. 2021; Basyuni et al. 2022b). Community participation has been crucial to successful rehabilitation efforts, such as in Tongke Tongke and South Sulawesi, where the local community initiates and maintains mangrove plantations (Amri 2005). To improve the success rate of mangrove rehabilitation, better approaches are needed, including proper land preparation, regular maintenance, and monitoring (Suyadi et al. 2023). Considering both short- and long-term economic benefits for local communities may enhance the sustainability of mangrove conservation efforts (Amri 2005) (Table 2).

2.3 *The Lack of Strict Law Enforcement Exacerbates Habitat Loss*

Owing to several factors, the lack of strict law enforcement exacerbates the habitat loss of mangroves in Indonesia. Among them are the weak implementation of environmental laws, corruption, and a lack of political will, all of which hinder effective enforcement (Zahroh and Najicha 2022). Overlapping authority claims among governmental institutions and conflicting regulations related to coastal reclamation,

Table 2 Examples of thriving mangrove plantations and rehabilitation in Indonesia

Sites	Criteria and indicators	Description site	References
Lubuk Kertang village, Langkat, North Sumatra	Increased species biodiversity (from 1 to 13–14 species), more than 80% seedling growth	Ex-pond	Amelia et al. (2023)
Four coastal villages (Sriwulan, Bedono, Timbulsloko, and Surodadi), Demak, Central Java, Indonesia	The CBMM applied in Bedono was most effective in maintaining the integrity of the rehabilitated ecosystem	Pond and riverbank	Damastuti and de Groot (2017)
Brebes, Central Java	Naturally, regenerated mangroves have shown better adaptation to environmental situations than transplanted mangroves	Coastlines	Nguyen et al. (2017)
Bays of South China Sea (Penibung, Sungai Duri, Karimunting), West Kalimantan	The effects of breakwaters on mangrove rehabilitation on the colonization rate of *Avicennia marina* and reforestation of *Rhizophora* sp.	Shoreline	Akbar et al. (2017)
Fishing villages of Kurri Caddi and Kurri Lompo, South Sulawesi	Increasing the establishment and survival rates of recruited vegetation and seedling establishment. The rehabilitation of appropriate surface elevations	Ponds	Oh et al. (2017)
Mahakam Delta in East Kalimantan, Indonesia	The complexity of managing dynamic coastal systems subjected to rapid human-induced biophysical changes exacerbated by a nonconducive policy environment aggravating the impact on the ground	Ponds	Powell and Osbeck (2010)
Tanakeke and Tiwoho, Sulawesi	The complexity of managing dynamic coastal systems subjected to rapid human-induced biophysical changes exacerbated by a nonconducive policy environment aggravating the impact on the ground	(Tiwoho, deep muds, and silty substrates) and low (Tanakeke, shallow, coralline sands) productivity mangrove ecosystems	Cameron et al. (2019)
Simeulue Island and Singkil lagoon, Aceh, Indonesia	Observation and monitoring of natural regeneration and calculation of rates of recruitment after a significant disturbance event is equally or more important than mangrove planting from not only ecological but also social and economic points of view	Tsunami events	Brown et al. (2015)

CBMM community-based mangrove management

mangrove management in non-forest zones, and aquaculture expansion contribute to ecosystem degradation (Mursyid et al. 2021). The loosening of licensing by local authorities for settlements and environmentally unfriendly exploitation further threatens mangrove forests (Buchwinkler 2022). To address these challenges, improved orchestration of policy priorities across institutions can strengthen political commitment, and enhanced law enforcement is imperative (Rivera-Monroy et al. 2022). Involving local communities in mangrove protection and conservation efforts is essential for sustainable management (Basyuni and Wati 2017; Mursyid et al. 2021; Rivera-Monroy et al. 2022). The implementation of these strategies is vital for supporting Indonesia's climate-change mitigation goals and preserving valuable mangrove ecosystems.

2.4 *Impact of Climate Change*

2.4.1 Rising Sea Levels and Increasing Frequency of Extreme Weather Events Threaten the Survival of Mangrove Plantations

One of the main threats facing mangrove forests is sea level rise (SLR), a phenomenon caused by global warming and the melting of polar ice caps. According to projections by the Intergovernmental Panel on Climate Change (IPCC), global sea levels could rise by up to 1 m by the end of the twenty-first century if greenhouse gas emissions remain unchecked (Bijsterveldt et al. 2023). This increase has affected all coastal ecosystems, including mangroves, through increased salinity, abrasion, and persistent inundation. Mangroves that cannot adapt quickly to these changes lose their habitat, especially if inland migration is blocked by human activities or infrastructure (Setyaningsih et al. 2019).

In Indonesia, this threat is exacerbated by subsidence caused by groundwater exploitation and developmental activities in coastal areas. In areas such as Semarang and Demak, subsidence rates reach 8–20 cm per year, accelerating the effects of relative sea level rise (RSLR) (Nurhidayah et al. 2022). Subsidence accelerates inundation and increases coastal abrasion, ultimately causing lateral retreat of mangrove forests. This affects the sustainability of the mangrove ecosystem as a natural stronghold and the coastal communities that depend on it (Table 3).

One approach to address the threat of rising sea levels in mangrove forests is vulnerability-based ecosystem restoration. On the basis of research on the northern coast of Java, Indonesia, priority areas for mangrove restoration can be identified via the coastal vulnerability index (CVI) and mangrove vulnerability index (MVI). The Demak region is the most vulnerable region due to intensive abrasion and mangrove loss. In contrast, Banyuwangi has low vulnerability due to its steep topography, which protects this region from erosion (Sagala et al. 2024). This approach provides a scientific basis for planning restorations that require immediate intervention.

Table 3 Impacts of sea-level rise (SLR) on mangroves in Indonesia

Location	SLR per year	Impact on mangroves	Proposed solutions	References
Mendahara Ilir, Sumatra	5–7 mm	Mangrove migration impeded; threat of inundation and erosion	Protection of migration pathways to inland areas	Setyaningsih et al. (2019)
Semarang & Demak	8–20 cm	Lateral retreat of mangroves; population migration	Groundwater resource management	Bijsterveldt et al. (2023)
Semarang & Demak	3.1–12 cm	Ineffective governance; reliance on large-scale infrastructure	Nature-based solutions and hybrid governance	Nurhidayah et al. (2022)
Northern Coast of Java	5 mm	High vulnerability in Demak; mangrove loss due to erosion	Restoration using CVI and MVI indices	Sagala et al. (2024)

CVI coastal vulnerability index, *MVI* mangrove vulnerability index

Ecosystem theory suggests that mangrove forests have adaptation mechanisms, such as root growth and sediment accretion, to environmental changes. However, this adaptability has limits. For example, research in Mendahara Ilir, Sumatra, revealed that natural sediment accretion rates only reach 2.2 mm per year, which is far below the modern SLR rates of 5–7 mm per year (Setyaningsih et al. 2019). Without effective intervention, mangrove forests are at risk of irreversible damage. In addition, poorly coordinated governance approaches at the government, sector, and stakeholder levels exacerbate this challenge.

The adaptability of mangroves to rising sea levels is strongly influenced by their sedimentation rate and their ability to migrate to relatively high land areas. Studies in Gazi Bay, Kenya, have shown that the ability of mangroves to migrate inland is strongly influenced by the availability of barrier-free spaces, such as the absence of human infrastructure, that impede this movement. However, if mangroves cannot migrate, these ecosystems are at risk of irreversible damage due to changes in the hydroperiod and increased inundation frequency (Nitto et al. 2014). These results emphasize the importance of spatial management in supporting natural ecosystem adaptation, including the protection of rear mangrove areas from development. Prominent infrastructure-based approaches such as levee construction are often favored over nature-based solutions such as mangrove restoration. Natural solutions are more sustainable and offer wider social and ecological benefits (Nurhidayah et al. 2022).

2.4.2 Changes in Salinity and Sedimentation Dynamics Hinder Seedling Establishment

Changes in salinity and sedimentation dynamics are major challenges for mangrove planting in Indonesia, as climate change increases the pressure on coastal ecosystems. Sedimentation dynamics, such as accretion (sediment deposition) and erosion (sediment erosion), can cause failure in mangrove seedling establishment. Accretion can totally or partially bury seedlings, causing etiolation or even death; erosion can expose the roots, reduce stability, and increase the risk of seedlings being uprooted by waves or currents. Both of these processes are substantial obstacles to successful mangrove restoration in dynamic coastal areas (Balke et al. 2013; Hespen et al. 2022).

High soil salinity is another major limiting factor in mangrove seedling establishment. Studies using individual-based models have shown that salinity affects mangrove productivity by reducing stomatal conductance and photosynthetic efficiency. At moderate salinities, species such as *Bruguiera gymnorhiza* grow well and dominate the habitat. However, when salinity exceeds a certain threshold (>30‰), the productivity of these species decreases dramatically, and more tolerant species, such as *Rhizophora stylosa*, become dominant. This emphasizes the importance of selecting mangrove species that are suitable for local salinity conditions to improve restoration success (Yoshikai et al. 2022).

Sedimentation dynamics also affect the stability of mangrove seedlings on the basis of their morphological characteristics. Species with large and long propagules, such as *R. stylosa*, are more tolerant to accretion than pioneer species, such as *Avicennia alba*. Pioneer species can adapt faster to erosion conditions by growing longer roots, thereby increasing their stability (Hespen et al. 2022). These adaptation strategies reflect important differences in the ability of different species to survive in coastal zones with dynamic sedimentation conditions.

Additional challenges arise from the impacts of climate change, such as increased storm intensity and sea level rise, which accelerate changes in sedimentation and salinity in coastal habitats. Persistent mechanical disturbances caused by sediment dynamics reduce the chances of mangrove seedlings reaching the stability phase. The successful natural colonization of seedlings often depends on short periods of undisturbed sedimentation that allow the seedlings to take root and develop. These conditions are often difficult to achieve in coastal environments that are highly disturbed by human activities and climate change (Balke et al. 2013, 2021).

To address these challenges, mangrove restoration efforts in Indonesia must integrate ecosystem-based approaches and technology. For example, temporary physical protection of seedlings during early growth stages can reduce the impact of erosion and accretion. Selecting mangrove species suitable for local salinity conditions and sedimentation dynamics is crucial for increasing success rates. Using this approach, mangrove restoration can protect coasts from the impacts of climate change and support ecosystem sustainability and the livelihood of coastal communities (Balke et al. 2013; Hespen et al. 2022; Yoshikai et al. 2022).

2.5 *Socioeconomic Constraints*

2.5.1 Limited Community Awareness and Financial Resources Impede Plantation and Conservation Efforts

Multilateral policy instruments not only support but also substantially influence community-based conservation. In pursuing biodiversity conservation and human welfare, community-based conservation includes the economic, community, and environmental aspects of communities living in the vicinity of forests. These connections will continue to foster forestry conservation in the future, ensuring that their objectives remain symbiotic. People cannot live independently of the environment and natural resources. Communities play an important role in forest maintenance and conservation. Therefore, conservation efforts, such as planting and restoring mangrove forests, must involve community education concerning the importance of protecting forests (Esmail et al. 2023).

Forest conservation during rehabilitation is challenging. However, it substantially affects the economic and social aspects of a community and can improve living standards. One of the reasons communities value conservation efforts is that they are able to derive the products of conservation, in this case, forestry products. Communities face diverse challenges in managing forest resources; limited access to basic facilities and infrastructure and a lack of financial resources exacerbate this management (Muttaqin et al. 2019).

The commitment and participation of local communities are key to the success of tropical mangrove plantation establishment. However, there is little awareness of the ecological and economic benefits of mangrove ecosystems in many Indonesian communities. Consequently, they often do not support plantation or conservation activities. In addition, the lack of income is challenging. A low income is a significant obstacle. Owing to insufficient funding for mangrove rehabilitation programs, local governments and communities face challenges regarding optimal implementation (Dale et al. 2014).

Bureaucracy and few viable opportunities for public–private partnerships are also hindrances to financing. Small-scale mangrove efforts are often not financially sustainable. Indonesia's mangrove conservation framework has neither fully integrated innovative financing mechanisms (e.g., carbon credits and ecotourism revenues) nor entirely encouraged their uptake. However, such financing mechanisms are prominent in the global literature (Choudhary et al. 2024).

2.6 *Conflicts of Interest Between Stakeholders, Such as Local Communities and Developers, Challenge Sustainable Management*

Stakeholder conflicts of interest constitute a considerable challenge to sustainable mangrove management in Indonesia. Local communities are sometimes (and understandably) tempted by short-term economic opportunities to generate income via activities such as aquaculture and timber extraction in mangroves, which they rely on for their livelihoods. For example, projects that require the intersection of neighboring states are often favored by developers, where regulations regarding infrastructure generally include urban or industrialization activities, which may include the removal of mangrove forests (Velde et al. 2019).

Cooperation becomes much more difficult for parties with conflicts of interest, where conservation agendas are fragmented and uncoordinated, and where dialog is avoided. Policies that allow the equitable sharing of benefits and inclusion in decision-making are imperative to mitigate these conflicts (Cormier-Salem 2006; Akram et al. 2023).

2.7 *Technical Limitations*

2.7.1 Inadequate Understanding of Species-Specific Requirements and Hydrological Needs Affects the Success of Plantations

Mangrove restoration efforts in Indonesia often face technical challenges, mainly because of a lack of understanding of species-specific requirements and appropriate hydrological conditions. This affects the success rate of mangrove restoration.

Species-Specific and Hydrological Needs

A lack of understanding of the specific needs of mangrove species is often the cause of failure in tropical mangrove restoration. Van Loon et al. (2016) reported that restoration often fails because the hydrological conditions do not match species-specific requirements. For example, the duration and frequency of inundation substantially influence the distributions of *Rhizophora* sp. and *Avicennia* sp. Understanding these patterns can lead to mangroves being planted at sites with prolonged inundation, which will not support long-term growth. Su et al. (2022) reported marked variations in the areas suitable for each species. *Avicennia marina* has been most extensively restored (approximately 4360 ha). This is due to its tolerance to high salinity and various environmental conditions (Aznawi et al. 2024). *Sonneratia apetala* is an exotic species often used for restoration due to its high adaptability to harsh environments, including its tolerance to low temperatures. *Aegiceras corniculatum* and *Acanthus ilicifolius* had the lowest restoration areas (~120 ha each) because of their sensitivity to specific environmental conditions.

Effect of Lack of Specific Data

Gillerot et al. (2018) emphasized the importance of species-specific data, such as wood density and carbon content, to ensure the success of restoration projects. Specific variations between and within mangrove species affect their adaptation to environmental conditions, such as salinity and elevation. Without such data, planting is often based on generalizations, which can increase restoration failure rates. Monoculture-based restoration (the widespread planting of one species) tends to fail because monocultures are more susceptible to pest attacks. The lack of species diversity limits the ability of an ecosystem to provide a range of ecosystem services (such as habitats for fish, birds, and other organisms). Although tolerant of a wide range of conditions, monoculture planting of *A. marina* can preclude the biodiversity benefits of mixed restoration (Su et al. 2022).

Species Adaptation to Substrate

Kasim (2022) reported that substrate adaptation and the specific morphological requirements of species (such as *A. lanata*) play important roles in restoration success. The mangrove forest requires specific substrates, such as mud and sandy soil, for optimal growth. A lack of understanding of these requirements has led to failures in restoration programs in Gorontalo, where substrate conditions do not meet the habitat requirements of the species (Basyuni et al. 2022b).

Hydrological Challenges in Indonesia

Indonesia faces unique challenges as an archipelago with marked hydrological variation. Van Loon et al. (2016) indicated that hydrological improvements, such as the removal of water flow barriers, can increase restoration success. However, projects often fail because of a lack of proper initial assessments of site hydrological conditions. The implementation of hydrological classification can address this issue. Mangrove restoration faces substantial hydrological challenges that can impede the success of rehabilitation. The key issues include altered tidal regimes, sediment dynamics, and water salinity levels, all of which are crucial to mangrove health and regeneration. Balke et al. (2021) developed innovative tools to monitor and address these challenges. One such tool was the “Mini Buoy,” which was designed to measure tidal hydrology in coastal wetlands. This device provides precise data on water levels and tidal patterns, which are essential for understanding the hydrological conditions necessary for successful mangrove restoration. By categorizing areas on the basis of their hydrological characteristics, restoration efforts can be tailored to the specific needs of each site, thereby increasing the likelihood of success.

2.7.2 Limited Availability of Quality Propagules and Seeds

The limited availability of high-quality seedlings and seeds is one of Indonesia’s main challenges in mangrove planting and rehabilitation. One of the main factors is the difficulty in obtaining propagules or seeds of superior quality in large quantities. Mangrove propagules have high mortality rates when faced with crab predation, strong water flows, and extreme environmental conditions. For example, research on *Rhizophora apiculata* has shown that cut propagules are more susceptible to strong currents and pests (Usman et al. 2022). Reliance on quality propagules requires increased production capacity in nurseries to meet large-scale rehabilitation needs.

Environmental quality issues around nurseries affect the outcomes of mangrove nurseries. Suboptimal environments, such as those with extreme soil pH or low organic matter content, can inhibit propagule growth. These environmental factors are exacerbated by the lack of attention given to the integration of nursery management with local ecosystems (Arifanti et al. 2022). As a result, the seedlings produced do not always survive or adapt well when they are planted at rehabilitation sites.

The limitations in nursery technology are a serious obstacle. Although methods such as tissue culture and vegetative propagation have been applied to some species, their effectiveness in supporting mass demand is still limited. Research in Indonesia has shown that vegetative propagation via whole or cut propagules requires sterile growing media and growth regulators to improve seedling quality (Arifanti et al. 2022; Usman et al. 2022). This adds to the cost and time burden, which is an obstacle to implementation in the field (Aznawi et al. 2024).

The distribution of natural propagules is an important issue. In many cases, the planting areas are far from natural propagule sources. This requires manual collection and distribution of propagules, which is labor- and resource-intensive. Approaches such as hydrological restoration have been attempted to facilitate natural propagule recruitment, but success varies depending on local conditions (Vanderklift et al. 2020).

2.7.3 Opportunities in Tropical Mangrove Plantations in Indonesia

The findings of the FAO (2023) emphasize the need to comprehend and forecast the determinants of natural mangrove formation to enhance restoration methods. Ecological restoration frameworks must account for natural colonization and succession processes, necessitating an understanding of the physical factors influencing the establishment of pioneer species and the adaptation of plant features to their local habitat.

2.8 Climate Change Mitigation

2.8.1 Mangrove Plantations Can Act as Important Carbon Sinks and Contribute to Indonesia's Nationally Determined Contributions (NDCs)

Mangroves are well known as among the most effective natural carbon sinks and can store four times as much carbon per hectare as terrestrial forests. These ecosystems are crucial buffers against climate change because they sequester carbon in their biomass and sediments. Indonesia, which has the most extensive global mangrove coverage, has incorporated mangroves into its NDCs by placing them as an important goal for reducing emissions (Dinilhuda et al. 2020).

Mangrove forests in Indonesia cover ~24% of the world's mangrove area (Arifanti et al. 2022). Globally, mangrove ecosystems are estimated to sequester 4.19 petagrams (Pgs) of carbon, a value heavily influenced by Indonesia (Hamilton and Friess 2016). Although mangrove forests cover only approximately 2.6% of Indonesia's overall forested area, degraded and deforested mangroves contribute to approximately 10% of carbon emissions in the forestry sector (Arifanti et al. 2022).

The conservation of existing mangrove forests is more effective than restoration in reducing emissions and, as such, is critical for the country meeting its NDCs (nationally determined contributions) under the Paris Agreement (Arifanti et al. 2022). Among Indonesia's natural climate solutions, the greatest mitigation potential is associated with mangroves and peatlands, where mangroves alone contribute an estimated 41.1 ± 1.4 megatons of CO_2e per year (Arifanti et al. 2021). Despite the positive intentions behind these efforts, their implementation depends on the establishment of firm legal proposals, as well as the active participation of all stakeholders, especially local communities (Tosun 2000).

Mangrove plantations offer many benefits in addition to carbon sequestration. These include improved coastal protection and biodiversity preservation, both of which help coastal communities become more resilient to climate change. These communities, which are frequently at the front lines of climate change, are also protected from the impacts of sea level rise and extreme weather events through the mitigation function of mangroves (Donato et al. 2011).

2.8.2 Restoration Programs Such as "Mangrove for the Future" Align with Global Carbon-Offset Markets

Mangrove restoration programs are increasingly viewed as promising candidates for carbon offsetting (Dinilhuda et al. 2020). Mangroves are crucial for climate change mitigation because of their ability to store carbon effectively and their role in nitrogen cycling (Mubaraq et al. 2024). These qualities make mangrove ecosystems promising participants in carbon markets, with the potential for massive restoration scaling. However, carbon removal must be real, additional, and permanent for the system of carbon offsets to be effective, and reforms are needed (Galatowitsch 2009).

Many methodologies have been approved for coastal wetland carbon offsets; however, the implementation of such projects is still limited, particularly in the United States (Sapkota and White 2019). The success of mangrove restoration often depends on our understanding of agroecological systems, hydrological patterns, and barriers to natural regeneration, which tend to be unique to the site. Community-based approaches that consider social and economic needs are recognized as potentially critical elements in climate adaptation through restoration (Archer et al. 2014). Although nascent, the market for mangrove carbon offsets offers considerable potential for the growth of global carbon trading and the fulfillment of hectare-for-hectare climate mitigation objectives (Sapkota and White 2019).

The future of mangroves has highlighted the potential of using mangrove plantations as methane minimizers to mitigate climate change. These initiatives aim to restore nonfunctioning mangrove sites and present opportunities to participate in emerging carbon-offset markets. International programs for generating verified carbon credits align with frameworks that complement those of the United Nations REDD+ program and represent sustainable forest management practices (Friess et al. 2022).

Mangrove plantations are efficient carbon-trading mechanisms that ensure high financial sustainability for restoration activities. Carbon-sink status encourages partnerships among local communities, government agencies, and private investors to ensure a fair distribution of proceeds from carbon markets (Howard et al. 2015). New synergies between these restoration efforts and not-for-profit and commercial carbon markets may be leveraged to accelerate and/or scale multiscale mangrove conservation implementations that benefit local economies and global climate targets.

2.9 Coastal Protection

With the world's longest coastline and tropical climate, Indonesia has an excellent opportunity to develop mangrove plantations as a sustainable coastal protection solution. Tropical mangroves can dampen wave energy owing to their dense root and stem structures, thereby creating effective barriers to water flow and wave energy. Compared with those without vegetation, beaches with mangroves can reduce erosion by up to 15 times (McIvor et al. 2012; Sanchez-Nunez et al. 2019). The geography of Indonesia, which is prone to high waves and storms, makes mangrove plantations a strategic opportunity for mitigating wave impacts. Indonesia's tropical mangroves can protect coastal areas from storm surges that often hit the tropics. Mangroves can reduce stormwater levels by 5–50 cm per kilometer of forest width, depending on the vegetation type and density (McIvor et al. 2012). As Indonesia experiences frequent tropical cyclones and high waves in some areas, mangrove plantations can be a natural solution for protecting coastal communities from economic loss and loss of life due to disasters (Blankespoor et al. 2017).

Mangrove plantations also offer ecological benefits for sediment stabilization and the restoration of degraded coastal ecosystems. In low-energy environments, mangroves support the sedimentation of fine particles, such as clay and silt, creating a stable and fertile soil layer (Blankespoor et al. 2017). This is particularly relevant in Indonesia, which often faces coastal damage due to the clearing of mangrove forests for ponds or development. The development of mangrove plantations in coastal areas threatened by damage can be an important step toward restoring the coastal environment.

Mangrove plantations provide excellent opportunities for improving community welfare. In addition to their coastal protection functions, mangroves provide a variety of resources, such as timber, fish, and shellfish, that support the livelihoods of local communities. In Thailand, the economic value of mangrove coastal protection services has reached $3679 per hectare per year (Sanchez-Nunez et al. 2019). If managed sustainably, mangrove plantations in Indonesia could become a new economic sector that supports the livelihoods of coastal communities while protecting ecosystems.

However, the successful development of mangrove plantations in Indonesia requires attention to climate change threats, such as rising sea levels, which may limit the space available for mangroves to migrate inland. Therefore, adaptation strategies such as mangrove restoration, reforestation, and limiting overexploitation are essential. Well-managed mangrove plantations protect Indonesia's coasts, increase the capacity of communities to adapt to climate change, and support the sustainability of coastal ecosystems (McIvor et al. 2012).

2.10 Plantation Programs Can Reduce the Cost of Coastal Infrastructure Maintenance

Mangroves are important in protecting coastlines from erosion and flood damage. Dense mangrove roots absorb wave energy and slow the flow of water before they reach the shore. This function reduces the impact of large waves and storms that can damage coastal infrastructure, such as roads, settlements, and other public facilities. Mangroves also serve as a natural barrier against damage, providing additional protection for coastlines that often face pressures from human activities and climate change. In a global study, combining mangroves with embankments reduced embankment height requirements by up to 10 cm, directly reducing coastal protection damage and costs (Kumano et al. 2021).

Mangrove restoration programs have also been shown to reduce economic losses caused by floods. For example, in the Caribbean region, the natural protection provided by mangroves substantially reduces economic damage, especially in areas where infrastructure and populations are vulnerable to flooding. Positive benefit–cost ratios were found at most restoration sites, suggesting that investments in mangrove ecosystems generate favorable returns. Mangroves are important components of disaster risk adaptation strategies in coastal areas because of their ability to reduce infrastructure damage, economic disruption, and postdisaster recovery costs (Beck et al. 2022).

Cost efficiency is one of the main advantages of mangrove planting programs over complex infrastructure solutions, such as concrete embankments or seawalls. By planting mangroves, the need to build or increase the length and height of embankments can be reduced by up to 37% in some locations, such as Vietnam. Because mangroves act as additional barriers that dampen wave energy, construction costs are reduced, and the life of complex infrastructure is extended. With this combination, the benefit-to-cost ratio is 1.5 times greater than when only gray infrastructure is used, making mangroves a more cost-effective and sustainable solution (Kumano et al. 2021).

In the face of climate change threats, mangroves are a sustainable solution for protecting coastlines and supporting economic development. In Indonesia, mangroves provide ecosystem benefits of an average of USD 15,000 per hectare per year. These benefits include coastal protection, fishery support, carbon sequestration, and economic opportunities from tourism. In areas with more developed coastal infrastructure, such as Bali and Java, the coastal protection value of mangroves has reached USD 39,970 per ha per year. With restoration costs of approximately USD 3900 per ha, mangroves can provide long-term benefits that far exceed the initial investment. In Indonesia, 600,000 ha of mangrove rehabilitation are planned through 2024 as a strategic move to increase coastal resilience to sea level rise and other natural disasters (Sasmito et al. 2023). This approach also helps the country achieve blue economy goals, which focus on environmental sustainability and the well-being of coastal communities. With careful planning, mangrove conservation and restoration can be the keys to creating a resilient and prosperous coastal future (Zanten et al. 2021).

2.11 Livelihood Enhancement

Mangrove ecosystems support coastal livelihoods and protect communities from natural disasters. They provide essential resources for local communities, including shellfish, crabs, shrimp, and fish (Basyuni et al. 2021; Akram et al. 2023). Mangroves also act as physical barriers against tidal influences and tsunamis, although their effectiveness depends on forest quality and extent (Dahdouh-Guebas et al. 2005). In Bangladesh, mangroves offer various livelihood options and environmental support to coastal communities (Hossain et al. 2018). Conservation efforts such as local-scale mangrove restoration have successfully enhanced livelihoods (Idrus et al. 2019).

Stock enhancement strategies for mangrove-associated crab fisheries have demonstrated promising results, with recovery rates of up to 50% and yield increases of up to 46%. However, habitat restoration might be more effective for long-term stock recovery in open systems. A balanced approach that integrates hatchery release and habitat restoration programs is recommended for sustainable mangrove ecosystem management (Le Vay et al. 2008).

2.11.1 Sustainable Mangrove Plantations Support Fishing, Ecotourism, and Honey Production and Offer Alternative Income Sources for Coastal Communities

Sustainable mangrove plantations are key ecosystem restorations with the potential for climate mitigation, biodiversity conservation, and promising prospects for coastal community livelihoods. These plantations also support various other income-generating activities, such as fisheries, ecotourism, and honey production, which serve as a foundation for economic resilience in vulnerable coastal areas.

Augmenting Fisheries

Mangrove forests are important nurseries for many commercially significant fish and crustaceans. Therefore, sustainable forms of mangrove plantations will enable the sustained delivery of these services and improve fishery benefits for local communities (Alongi 2021; Basyuni et al. 2021). Better fish stocks bolster food security and provide additional income sources for coastal households that rely on fishing for their livelihood.

Ecotourism Development

Unlike other biomes that require extensive infrastructure, mangrove ecosystems are targeted explicitly for ecotourism opportunities that attract visitors through specialized offerings, such as birdwatching, kayaking, and guided tours. Community-based ecotourism projects return profits from nature-based tourism to the community (Spalding and Parrett 2019). These initiatives help build supplementary income and raise awareness among tourists and native people about the importance of mangrove conservation.

Honey Production

Because many mangrove species flower cyclically, they can contribute nectar as a constant source in plantations to support apiculture (beekeeping). Through honey production, coastal communities have another cash flow option and the opportunity to diversify (Basyuni et al. 2024). In addition, sustainable honey production from mangroves has been recognized as an eco-friendly activity, implying its potential for scaling up eco-friendly enterprises (Das et al. 2023).

Integrating Livelihoods into Conservation

This practice also plays a role in the long-term sustainability of these ecosystems by offering opportunities for mangrove restoration and conservation. New organically participatory management practices (in which local stakeholders are involved in planning and decision-making) show that there is a tradeoff between ecological goals and socioeconomic advantages (Friess et al. 2022).

2.12 *Policy Support and Funding Opportunities*

Global warming is a major contributing factor to climate change. High sea levels inundate various coastal areas, thereby affecting mangrove ecosystems. Mangroves have the highest potential for sequestering blue carbon. Forest conservation plays a crucial role in preserving the stability of coastal ecosystems. Climate change solutions are on the agenda and legitimized in scientific and political processes; therefore, science frames the global climate crisis (Pryck and Boettcher 2024).

Several international and nongovernmental organizations (NGOs) have recommended investigating conservative techniques to preserve mangrove and seagrass forests and salt marshes. However, despite the existence of programs aimed at reducing carbon emissions from coastal development, capitalizing on existing opportunities, and advancing our understanding of blue carbon (Donato et al. 2011), there is still a lack of basic economic research on the viability of carbon credit systems, such as Reducing Emissions from Deforestation and Forest Degradation (REDD) programs. Consequently, national strategies (such as mangrove restoration) can be important assets in the international carbon credit system; global attention to government policies can provide opportunities for international and nongovernmental organizations (NGOs) to benefit from this.

Indonesia has implemented ambitious mangrove rehabilitation programs to address coastal degradation and support sustainable development goals (Sasmito et al. 2023). The National Mangrove Rehabilitation Program aims to restore 600,000 ha by 2024, although a more realistic target of 193,367 ha has been identified on the basis of geomorphological factors and land use changes (Sasmito et al. 2023). These efforts involve community groups and provide economic and social benefits while increasing awareness of climate change (Ariyanto and Pringgenies 2024). Various planting designs and techniques have been used, with *Rhizophora* spp. commonly used for rehabilitation (Kusmana 2017). However, challenges persist, including low success rates and small-scale applications (Sasmito et al. 2023). Strategies for sustainable mangrove management, such as improving ecosystem protection, strengthening political commitment, involving stakeholders, and advancing research, are crucial (Rivera-Monroy et al. 2022). Successful implementation could benefit 74 million coastal people and contribute substantially to national emission reduction targets (Sasmito et al. 2023).

3 Ecological Interaction of Mangrove and *Eucalyptus* Plantations

Studies in Indonesia have examined industrial forestry operations in which the replacement of *Acacia mangium* with *Eucalyptus pellita* was necessitated by disease. Hardiyanto et al. (2024) and Nambiar et al. (2018) reported that this forced-species transition reduced growth rates and decreased plantation productivity. In East Kalimantan, Mackensen et al. (2003) reported that nutrient loss from industrial tree plantations compromised soil fertility. Prihatin and Wicaksono (2020) described land ownership conflicts between indigenous communities and state forestry claims. Abram et al. (2014) reported that the conversion of grasslands and swamp forests into palm oil plantations created conflicts among conservation goals, local livelihoods, and economic development.

However, few *Eucalyptus* studies have directly assessed their interactions with mangrove ecosystems. The current research focuses on plantation forestry and land-use changes; accordingly, there is no evidence of potential conflict between *Eucalyptus* plantations and mangroves in Indonesia.

Eucalyptus plantations in coastal regions alter local water dynamics and nutrient cycles, often with implications for adjacent ecosystems, such as mangroves. In *Eucalyptus*-dominated settings, evapotranspiration exceeded 80% of precipitation in 71% of observations, and streamflow accounted for only 5–11% of precipitation in stands aged 1–7 years (Balieiro et al. 2007). A total of 33.4% of the total rainfall was intercepted by tree trunks; shifts in the quality of organic material entering water systems were linked to altered nutrient dynamics, including measured contributions of 8.42 $kg.ha^{-1}$ ammonium nitrogen and 19.04 $kg.ha^{-1}$ potassium via canopy processes (Balieiro et al. 2007).

However, in mangrove habitats, research on plant invasion has reported that nonnative species can outcompete native flora and disrupt ecosystem functions. Similar concerns arise when *Eucalyptus* plantations transition near coastal and mangrove ecosystems, where changes in water quality, reduced water availability, and modified nutrient flow may affect biodiversity and ecosystem services. Management strategies—such as site-specific forest practices, invasive species control, and frameworks that balance wood production with water resource allocation—address these hydrological and ecological challenges (Augusthy et al. 2024).

4 Concluding Remarks

The growth of tropical mangrove plantations in Indonesia presents formidable obstacles but offers exciting prospects. Some major obstacles include the intricacies of species adaptation, site selection, hydrological needs, disease and pest control, and local socioeconomic dynamics. Opportunities to improve coastal protection, repair damaged ecosystems, back blue-carbon projects, and create jobs through sustainable mangrove-based industries, however, more than offset these difficulties. To address these issues, an integrated strategy that incorporates community involvement, scientific research, and policy assistance is needed. Emphasizing ecological compatibility, biodiversity conservation, and socioeconomic benefits can guarantee the long-term viability of mangrove plantations. As leaders in mangrove ecosystems worldwide, Indonesia can establish standards for the management of tropical mangrove plantations that promote environmental and economic sustainability while reducing the effects of climate change and enhancing coastal resilience. Industrial forestry operations in Indonesia have transitioned from *Acacia mangium* to *Eucalyptus pellita* due to disease, leading to reduced growth rates and productivity challenges. There are few direct studies on *Eucalyptus* and mangrove interactions, but evidence suggests that *Eucalyptus* plantations substantially alter local hydrology and nutrient cycles, potentially affecting adjacent mangrove ecosystems. High water consumption by *Eucalyptus* can reduce water availability for mangroves, thereby affecting their health and biodiversity. To mitigate these effects, it is essential to implement site-specific forest management practices, control invasive species, and balance wood production with water resource conservation.

Potential Conflicts of Interest The authors declare that they have no conflicts of interest.

Funding This work was supported by the e-ASIA Joint Program from the Endowment Fund for Education Agency (Lembaga Pengelola Dana Pendidikan), Ministry of Finance, Republic of Indonesia, year 2024–2025 (019/E5/PG.02.00/PRPB BATCH 2/2024).

References

Abram NK, Xofis P, Tzanopoulos J, MacMillan DC, Ancrenaz M, Chung R et al (2014) Synergies for improving oil palm production and forest conservation in floodplain landscapes. PLoS One 9(6):e95388

Akbar AA, Sartohadi J, Djohan TS, Ritohardoyo S (2017) The role of breakwaters on the rehabilitation of coastal and mangrove forests in West Kalimantan, Indonesia. Ocean Coastal Manag 138:50–59

Akram H, Hussain S, Mazumdar P, Chua KO, Butt TE, Harikrishna JA (2023) Mangrove health: a review of functions, threats, and challenges associated with mangrove management practices. Forests 14(9):1698. https://doi.org/10.3390/f14091698

Alongi DM (2012) Carbon sequestration in mangrove forests. Carbon Manag 3(3):313–322

Alongi DM (2021) Macro and micronutrient cycling and crucial linkages to geochemical processes in mangrove ecosystems. J Mar Sci Eng 9(5):456

Amelia R, Basyuni M, Alfinsyahri A, Sulistiyono N, Slamet B, Bimantara Y et al (2023) Evaluation of plant growth and potential of carbon storage in the restored mangrove of an abandoned pond in Lubuk Kertang, North Sumatra, Indonesia. Forests 14(1):158

Amri A (2005) Mangrove plantation and land property rights: a lesson from the coastal area of South Sulawesi, Indonesia. Japanese. J Southeast Asian Stud 43(2):141–160

Archer D, Almansi F, DiGregorio M, Roberts D, Sharma D, Syam D (2014) Moving toward inclusive urban adaptation: approaches to integrating community-based adaptation to climate change at city and national scale. Clim Dev 6(4):345–356

Ardianto H, Alfirdaus L, Manalu S, Kushandajani K (2020) Ecological destruction in Indonesia: Reflecting on the experience of East Kalimantan and Bangka Belitung. In: Proceedings of the 4th International Conference on Indonesian Social and Political Enquiries, ICISPE 2019, 21–22 October 2019. Semarang, Central Java, Indonesia

Arifanti VB (2020) Mangrove management and climate change: a review in Indonesia. IOP Conf Ser Earth Environ Sci 487:012022

Arifanti VB, Novita N, Tosiani A (2021, October) Mangrove deforestation and CO_2 emissions in Indonesia. IOP Conf Ser Earth Environ Sci 874:012006

Arifanti VB, Sidik F, Mulyanto B, Susilowati A, Wahyuni T, Yuniarti N, Novita N (2022) Challenges and strategies for sustainable mangrove management in Indonesia: a review. Forests 13(5):695

Ariyanto D, Pringgenies D (2024) Nutrient distribution in natural and rehabilitated mangrove with various types of mangrove, Lampung, Indonesia. Int J Design Nat Ecodyn 19(6):2171–2178

Asbridge E, Lucas R, Ticehurst C, Bunting P (2016) Mangrove response to environmental change in Australia's Gulf of Carpentaria. Ecol Indic 73:681–698

Augusthy S, Nizam A, Kumar A (2024) The diversity, drivers, consequences and management of plant invasions in the mangrove ecosystems. Sci Total Environ 173851

Aznawi AA, Basyuni M, Hanafiah DS, Larekeng SH, Mulya Sari SP (2024) Impact of pest and diseases on mangrove forest rehabilitation in Indonesia: a review. OnLine J Biol Sci 24(4):728–738

Balieiro FDC, Franco AA, Fontes RLF, Dias LE, Campello EFC, Faria SMD (2007) Evaluation of the throughfall and stemflow nutrient contents in mixed and pure plantations of Acacia mangium, Pseudosamenea guachapele and Eucalyptus grandis. Revista Árvore 31:339–346

Balke T, Webb EL, van den Elzen E, Galli D, Herman PM, Bouma TJ (2013) Seedling establishment in a dynamic sedimentary environment: a conceptual framework using mangroves. J Appl Ecol 50(3):740–747

Balke T, Vovides A, Schwarz C, Chmura GL, Ladd C, Basyuni M (2021) Monitoring tidal hydrology in coastal wetlands with the "mini buoy": applications for mangrove restoration. Hydrol Earth Syst Sci 25(3):1229–1244

Basyuni M, Sulistiyono N (2018) Deforestation and reforestation analysis from land-use changes in North Sumatran Mangroves, 1990–2015. IOP Conf Ser Mater Sci Eng 309(1):012018. IOP Publishing

Basyuni M, Wati R (2017) Bioinformatics analysis of the oxidosqualene cyclase gene and the amino acid sequence in mangrove plants. J Phys Conf Ser 801(1):012011. IOP Publishing

Basyuni M, Slamet B, Sulistiyono N, Munir E, Vovides AG, Bunting P (2021) Physicochemical characteristic, nutrient, and fish production in different types of mangrove forest in North Sumatra and Aceh provinces of Indonesia. Kuwait J Sci 48(3)

Basyuni M, Sasmito SD, Analuddin K, Ulqodry TZ, Saragi-Sasmito MF, Eddy S, Milantara N (2022a) Mangrove biodiversity, conservation, and roles for livelihoods in Indonesia. In: Mangroves: biodiversity, livelihoods, and conservation. Singapore, Springer Nature Singapore, pp 397–445

Basyuni M, Bimantara Y, Cuc NT, Balke T, Vovides AG (2022b) Macrozoobenthic community assemblage as key indicator for mangrove restoration success in North Sumatra and Aceh, Indonesia. Restor Ecol 30(7):e13614

Basyuni M, Mubaraq A, Aznawi AA, Sivaipram I, Amir AA, Jumilawaty E et al (2024) Kelulut (*Trigona* spp.) cultivation to increase production and additional income of mangrove ecotourism Peduli Pesisir Farmer Groups, Beras Basah Village, Pangkalan Susu Sub-District, North Sumatra, Indonesia. J Comm Serv Sustain Empowerm 4(02):1–11

Basyuni M, Amelia R, Aznawi A, Wirasatriya A, Iryanthony S, Slamet B, Al Mustaniroh S, Rahmania R, Rahmila Y, Sumarga E, Larekeng S, Salmo S III, Kajita T, Sivaipram I, Ali H (2025) Reduction of mangrove carbon stock ecosystems due to illegal logging using a combination of unmanned aerial vehicle imagery and field surveys. Global J Environ Sci Manag 11(1):225–242

Beck MW, Heck N, Narayan S, Menéndez P, Reguero BG, Bitterwolf S et al (2022) Return on investment for mangrove and reef flood protection. Ecosyst Serv 56:101440
Bijsterveldt CE, Herman PM, van Wesenbeeck BK, Ramadhani S, Heuts TS, van Starrenburg C, Bouma TJ (2023) Subsidence reveals potential impacts of future sea level rise on inhabited mangrove coasts. Nat Sustainability 6(12):1565–1577
Blake WH, Kelly C, Wynants M, Patrick A, Lewin S, Lawson J, Ndakidemi P (2021) Integrating land-water-people connectivity concepts across disciplines for codesign of soil erosion solutions. Land Degrad Dev 32(12):3415–3430
Blankespoor B, Dasgupta S, Lange GM (2017) Mangroves as a protection from storm surges in a changing climate. Ambio 46(4):478–491
Bosma R, Sidik AS, van Zwieten P, Aditya A, Visser L (2012) Challenges of a transition to a sustainably managed shrimp culture agro-ecosystem in the Mahakam delta, East Kalimantan, Indonesia. Wetl Ecol Manag 20:89–99
Brown B, Yuniati W, Ahmad R, Soulsby I (2015) Observations of natural recruitment and human attempts at mangrove rehabilitation after seismic (tsunami and earthquake) events in Simeulue Island and Singkil Lagoon, Aceh, Indonesia. Reconstruction and restoration, Posttsunami hazard, pp 311–327
Buchwinkler DM (2022) Local impacts of the global shrimp industry: mangrove clearings, shrimp farms and environmental justice in Sabah, Malaysia. ÖFSE-Forum
Cahyaningsih AP, Deanova AK, Pristiawati CM, Ulumuddin YI, Kusumaningrum L, Setyawan AD (2022) Causes and impacts of anthropogenic activities on mangrove deforestation and degradation in Indonesia. Int J Bonorowo Wetlands 12(1)
Cameron C, Hutley LB, Friess DA, Brown B (2019) Community structure dynamics and carbon stock change of rehabilitated mangrove forests in Sulawesi, Indonesia. Ecol Appl 29(1):e01810
Choudhary B, Dhar V, Pawase AS (2024) Blue carbon and the role of mangroves in carbon sequestration: its mechanisms, estimation, human impacts and conservation strategies for economic incentives. J Sea Res 199:102504
Cormier-Salem MC (2006) Mangrove: changes and conflicts in claimed ownership, uses and purposes. In: Environment and livelihoods in tropical coastal zones: managing agriculture fishery aquaculture conflicts. CABI, Wallingford, UK, pp 163–176
Dahdouh-Guebas F, Jayatissa LP, Di Nitto D, Bosire J O, Seen DL and Koedam N (2005) How effective were mangroves as a defence against the recent tsunami?. Current Biol 15(12):R443–R447
Dale PER, Knight JM, Dwyer PG (2014) Mangrove rehabilitation: a review focusing on ecological and institutional issues. Wetl Ecol Manag 22:587–604
Damastuti E, de Groot R (2017) Effectiveness of community-based mangrove management for sustainable resource use and livelihood support: a case study of four villages in Central Java, Indonesia. J Environ Manag 203:510–521
Das A, Choudhury KM, Choudhury AK (2023) An assessment of mangrove vegetation changes in reference to cyclone-impacted climatic alterations at the land-ocean interface of the Indian Sundarbans with the application of remote sensing-based analytical tools. Environ Sci Pollut Res 30(38):89311–89335
Dinilhuda A, Akbar AA, Herawaty H (2020) Potentials of mangrove ecosystem as storage of carbon for global warming mitigation. Biodivers J Biol Divers 21(11)
Donato DC, Kauffman JB, Murdiyarso D, Purbopuspito J (2011) Mangroves among the most carbon-rich forests in the tropics per area. Nat Geosci 14(6):414–419. https://doi.org/10.1038/s41561-021-00749-z
Eddy S, Milantara N, Sasmito SD, Kajita T, Basyuni M (2021) Anthropogenic drivers of mangrove loss and associated carbon emissions in South Sumatra, Indonesia. *Forests* 12(2):187
Ellison AM (2008) Managing mangroves with benthic biodiversity in mind: moving beyond roving banditry. J Sea Res 59:2–15
Esmail N, McPherson JM, Abulu L, Amend T, Amit R, Bhatia S, Bikaba D, Colombi TAB, Brown J, Buschman V et al (2023) What's on the horizon for community-based conservation? Emerging threats and opportunities. Ecol Evol 38(7):666–680

FAO (2023) The world's mangroves 2000–2020. Food and Agriculture Organization of the United Nations
Fatoyinbo T, Feliciano EA, Lagomasino D, Lee SK, Trettin C (2018) Estimating mangrove aboveground biomass from airborne LiDAR data: a case study from the Zambezi River delta. Environ Res Lett 13(2):025012
Friess DA, Rogers K, Lovelock CE, Krauss KW, Hamilton SE, Lee SY, Shi S (2019) The state of the world's mangrove forests: past, present, and future. Annu Rev Environ Resour 44(1):89–115
Friess DA, Adame MF, Adams JB, Lovelock CE (2022) Mangrove forests under climate change in a 2C world. Wiley Interdiscip Rev Clim Chang 13(4):e792
Galatowitsch SM (2009) Carbon offsets as ecological restorations. Restor Ecol 17(5):563–570
Gillerot L, Vlaminck E, De Ryck DJ, Mwasaru DM, Beeckman H, Koedam N (2018) Inter-and intraspecific variation in mangrove carbon fraction and wood specific gravity in Gazi Bay, Kenya. Ecosphere 9(6):e02306
Giri C, Ochieng E, Tieszen LL, Zhu Z, Singh A, Loveland T, Duke N (2011) Status and distribution of mangrove forests in the world using earth observation satellite data. Glob Ecol Biogeogr 20(1):154–159
Giri S, Samanta S, Mondal PP, Basu O, Khorat S, Chanda A, Hazra S (2022) A geospatial assessment of the growth pattern of aquaculture in the Indian Sundarbans Biosphere Reserve. Environ Dev Sustain 24:4203–4225
Hakim L, Siswanto D, Makagoshi N (2017) Mangrove conservation in East Java: the ecotourism development perspectives. J Trop Life Sci 7(3):277–285
Hamilton S, Friess D (2016) High-resolution annual maps of whole system mangrove carbon stocks from 2000 to 2012. arXiv preprint arXiv:1611.00307, 26
Hardiyanto EB, Inail MA, Nambiar S, Mendham DS (2024) Sustaining plantation forest productivity in Sumatra over three decades: from acacias to eucalypts. For Ecol Manag 553:121613
Hespen R, Hu Z, Peng Y, Zhu Z, Ysebaert T, Bouma TJ (2022) Identifying trait-based tolerance to sediment dynamics during seedling establishment across eight mangrove species. Limnol Oceanogr 67(10):2281–2295
Hossain MA, Ahmed M, Ojea E, Fernandes JA (2018) Impacts and responses to environmental change in coastal livelihoods of South–West Bangladesh. Sci Total Environ 637:954–970
Howard RJ, Tallontire A, Stringer L, Marchant R (2015) Unraveling the notion of "fair carbon": key challenges for standards development. World Dev 70:343–356
Idrus AA, Syukur A, Zulkifli L (2019) The livelihoods of local communities: evidence success of mangrove conservation on the coastal of East Lombok Indonesia. AIP Conf Proc 2199(1) AIP Publishing
Kasim F (2022) Characteristics of *Avicennia lanata* (Ridley) species for mangrove restoration on the coast of North Gorontalo. Tomini J Aquat Sci 3(1):8–20
Kodikara KAS, Mukherjee N, Jayatissa LP, Dahdouh-Guebas F, Koedam N (2017) Have mangrove restoration projects worked? An in-depth study in Sri Lanka. Restor Ecol 25(5):705–716
Kumano N, Tamura M, Inoue T, Yokoki H (2021) Estimating the cost of coastal adaptation using mangrove forests against sea level rise. Coast Eng J 63(3):263–274
Kusmana C (2014) Distribution and current status of mangrove forests in Indonesia. In: Faridah-Hanum I, Latiff A, Hakeem K, Ozturk M (eds) Mangrove ecosystems of Asia. Springer, New York, pp 37–60
Kusmana C (2017) Lesson learned from mangrove rehabilitation program in Indonesia. Jurnal Pengelolaan Sumberdaya Alam dan Lingkungan (J Nat Res Environ Manag) 7(1):89–97
Le Vay L, Lebata MJH, Walton M, Primavera J, Quinitio E, Lavilla-Pitogo C, Wille M (2008) Approaches to stock enhancement in mangrove-associated crab fisheries. Rev Fish Sci 16(1–3):72–80
Lewis RR (2005) Ecological engineering for successful management and restoration of mangrove forests. Ecol Eng 24(4):403–418
Mackensen J, Klinge R, Ruhiyat D, Fölster H (2003) Assessment of management-dependent nutrient losses in tropical industrial tree plantations. AMBIO: A J Hum Environ 32(2):106–112

Maiti SK, Bandyopadhyay S, Mukhopadhyay S (2021) Importance of selection of plant species for successful ecological restoration programs in coal mine-degraded land. In: Phytorestoration of abandoned mining and oil drilling sites. Elsevier, pp 325–357

McIvor AL, Spencer T, Möller I, Spalding M (2012) Storm surge reduction by mangroves. The Nature Conservancy and Wetlands International

Meilianda E, Pradhan B, Comfort LK, Alfian D, Juanda R, Syahreza S, Munadi K (2019) Assessment of posttsunami disaster land use/land cover change and potential impact of future sea-level rise to low-lying coastal areas: a case study of Banda Aceh coast of Indonesia. Int J Disaster Risk Reduct 41:101292

Ministry of Environment and Forestry (2021) National Mangrove Ecosystem Management Strategy (2021–2030)

Mubaraq A, Sembiring M, Widiastuti E, Fachrial E, Utomo B, Turjaman M, Sidik F, Ulumuddin YI, Arifanti VB, Siregar ES, Kajita T, Procheş Ş, Basyuni M (2024) Diversity and distribution of nitrifying bacteria play an important role in the nitrogen cycle in mangrove sediments. Global J Environ Sci Manag 10(4):2145–2162

Mursyid H, Daulay MH, Pratama AA, Laraswati D, Novita N, Malik A, Maryudi A (2021) Governance issues related to the management and conservation of mangrove ecosystems to support climate change mitigation actions in Indonesia. Forest Policy Econ 133:102622

Muttaqin MZ, Alviya I, Lugina M, Hamdani FAU (2019) Developing community-based forest ecosystem service management to reduce emissions from deforestation and forest degradation. Forest Policy Econ 108:101938

Nambiar EKS, Harwood CE, Mendham DS (2018) Paths to sustainable wood supply to the pulp and paper industry in Indonesia after diseases have forced a change of species from acacia to eucalypts. Aust For 81(3):148–161

Nguyen TP, Luom TT, Parnell KE (2017) Mangrove transplantation in Brebes regency, Indonesia: lessons and recommendations. Ocean Coastal Manag 149:12–21

Nitto D, Neukermans G, Koedam N, Defever H, Pattyn F, Kairo JG, Dahdouh-Guebas F (2014) Mangroves facing climate change: landward migration potential in response to projected scenarios of sea level rise. Biogeosciences 11(3):857–871

Nurhidayah L, Davies P, Alam S, Saintilan N, Triyanti A (2022) Responding to sea level rise: challenges and opportunities to govern coastal adaptation strategies in Indonesia. Maritime Stud 21(3):339–352

Nursamsi I, Komala WR (2017) Assessment of the successfulness of mangrove plantation program through the use of open source software and freely available satellite. Nusantara Biosci 9(3):251–259

Obidzinski K (2003) Logging in East Kalimantan, Indonesia. The historical expedience of illegality (Thesis, fully internal). Universiteit van Amsterdam

Oh RRY, Friess DA, Brown BM (2017) The role of surface elevation in the rehabilitation of abandoned aquaculture ponds to mangrove forests, Sulawesi, Indonesia. Ecol Eng 100:325–334

Powell N, Osbeck M (2010) Approaches for understanding and embedding stakeholder realities in mangrove rehabilitation processes in Southeast Asia: lessons learnt from the Mahakam Delta, East Kalimantan. Sustain Develop 18(5):260–270

Prihatin PS, Wicaksono A (2020) The Conflict in Indonesia's Forestry Sector: A Case Study of a Community Conflict with Company Industrial Plantation Forest (HTI) in Pelalawan Riau Province. ICONEG 2019, October 25–26, Makassar, Indonesia

Primavera JH, Esteban JMA (2008) A review of mangrove rehabilitation in The Philippines: successes, failures, and future prospects. Wetl Ecol Manag 16:345–358

Pryck KD, Boettcher M (2024) The rise, fall and rebirth of ocean carbon sequestration as a climate 'solution'. Glob Environ Chang 85:102820

Quevedo JMD, Lukman KM, Ulumuddin YI, Uchiyama Y, Kohsaka R (2023) Applying the DPSIR framework to qualitatively assess the globally important mangrove ecosystems of Indonesia: a review toward evidence-based policymaking approaches. Mar Policy 147:105354

Richards DR, Friess DA (2016) Rates and drivers of mangrove deforestation in Southeast Asia, 2000–2012. Proc Natl Acad Sci 113(2):344–349

Rivera-Monroy VH, Zhao X, Wang H, Xue ZG (2022) Are existing modeling tools useful to evaluate outcomes in mangrove restoration and rehabilitation projects? A minireview. Forests 13(10):1638

Sagala PM, Bhomia RK, Murdiyarso D (2024) Assessment of coastal vulnerability to support mangrove restoration in the northern coast of Java, Indonesia. Reg Stud Mar Sci 70:103383

Sanchez-Nunez DA, Bernal G, Mancera Pineda JE (2019) The relative role of mangroves on wave erosion mitigation and sediment properties. Estuar Coasts 42:2124–2138

Sapkota Y, White JR (2019) Marsh edge erosion and associated carbon dynamics in coastal Louisiana: a proxy for future wetland-dominated coastlines worldwide. Estuar Coast Shelf Sci 226:106289

Sasmito SD, Basyuni M, Kridalaksana A, Saragi-Sasmito MF, Lovelock CE, Murdiyarso D (2023) Challenges and opportunities for achieving sustainable development goals through restoration of Indonesia's mangroves. Nat Ecol Evol 7(1):62–70

Setyaningsih CA, Biagioni S, Saad A, Kashima K, Sabiham S, Behling H (2019) Response of mangroves to late Holocene Sea-level change: paleoecological evidence from Sumatra, Indonesia. Wetlands 39:1103–1118

Slamet NS, Dargusch P, Aziz AA, Wadley D (2020) Mangrove vulnerability and potential carbon stock loss from land reclamation in Jakarta Bay, Indonesia. Ocean Coastal Manag 195:105283

Spalding MD, Leal M (eds) (2021) The state of the world's mangroves 2021. Global Mangrove Alliance

Spalding M, Parrett CL (2019) Global patterns in mangrove recreation and tourism. Mar Policy 110:103540

Su J, Yin B, Chen L, Gasparatos A (2022) Priority areas for mixed-species mangrove restoration: the suitable species in the right sites. Environ Res Lett 17(6):065001

Suyadi S, Nurdiansah D, Renyaan J, Hapsari BW, Rahayu EMD, Sugiharto A, Ulumuddin YI (2023) Better approaches are required for successful mangrove restoration and rehabilitation program. AIP Conf Proc 2606(1) AIP Publishing

Tosun C (2000) Limits to community participation in the tourism development process in developing countries. Tour Manag 21(6):613–633

Usman AHA, Hartoyo APP, Kusmana C (2022) The growth performance of *Rhizophora apiculata* using the cut-propagule method for mangrove rehabilitation in Indonesia. Biodivers J Biol Divers 23(12)

Utami W, Sugiyanto C, Rahardjo N (2024) Mangrove area degradation and management strategies in Indonesia: a review. J Degraded Mining Lands Manag 11(3)

Van Loon AF, Te Brake B, Van Huijgevoort MH, Dijksma R (2016) Hydrological classification, a practical tool for mangrove restoration. PLoS One 11(3):e0150302

Vanderklift MA, Doropoulos C, Gorman D, Leal I, Minne AJ, Statton J, Wernberg T (2020) Using propagules to restore coastal marine ecosystems. Front Mar Sci 7:724

Velde KV, Hugé J, Friess DA, Koedam N, Dahdouh-Guebas F (2019) Stakeholder discourses on urban mangrove conservation and management. Ocean Coastal Manag 178:104810

Yoshikai M, Nakamura T, Suwa R, Sharma S, Rollon R, Yasuoka J, Nadaoka K (2022) Predicting mangrove forest dynamics across a soil salinity gradient using an individual-based vegetation model linked with plant hydraulics. Biogeosciences 19(6):1813–1832

Zahroh UAZ, Najicha FU (2022) Problems and challenges on environmental law enforcement in Indonesia: AMDAL in the context of administrative law. Indonesian State Law Rev (ISLRev) 5(2):53–66

Zanten BT, Brander LM, Gutierrez Torres D, Uyttendaele GYP, Herrera Garcia LD, Patrama D, Kaczan DJ (2021) The economics of large-scale mangrove conservation and restoration in Indonesia. Econ Large-Scale Mangrove Conserv Restorat Indonesia 10:37605

Application of Near-Infrared (NIR) Spectroscopy for Measuring the Wood Quality of Plantation-Grown *Eucalyptus* Trees

Adlin Sabrina Muhamad Roseley and Noorsyazwani Mansoor

Abstract This chapter reviews recent studies on the application of near-infrared (NIR) spectroscopy for wood quality assessment. The NIR method has been proven to be a reliable, inexpensive, and rapid alternative to traditional destructive testing methods. A growing body of research has demonstrated that NIR spectroscopy is a powerful analytical tool for measuring the chemical, anatomical, physical, and mechanical properties of wood. The analytical capability of NIR spectroscopy is considered very useful in forest plantation settings, where rapid and easy application is an advantage. In the future, there will be more prospects for analytical studies on the effects of silvicultural treatment on wood quality with the advancement of NIR technology.

1 Introduction

Efforts to increase the timber supply in Southeast Asia have led to the use of plantation species such as *Eucalyptus* as an alternative source of raw material. Owing to their fast growth characteristics and valuable traits derived from interspecies hybridization for pulp and sawn timber, *Eucalyptus* is a promising option. Once the trees reach 10 years of age or older, the timber is often used in furniture or processed into peeler logs. In many countries, *Eucalyptus* has become one of the most extensively planted species for pulp production.

The properties of plantation-grown wood, such as *Eucalyptus*, are governed by genetic factors, silvicultural practices, and environmental conditions. Tree

A. S. Muhamad Roseley (✉)
Institute of Tropical Forestry and Forest Products, Universiti Putra Malaysia, Serdang, Selangor, Malaysia

Faculty of Forestry and Environment, Universiti Putra Malaysia, Serdang, Selangor, Malaysia
e-mail: adlin@upm.edu.my

N. Mansoor
Faculty of Forestry and Environment, Universiti Putra Malaysia, Serdang, Selangor, Malaysia

S. Mohd Razali et al. (eds.), *Sustainable Eucalyptus Forestry in Southeast Asia*,
Forestry Sciences 88, https://doi.org/10.1007/978-3-032-21343-3_6

improvement efforts have focused primarily on increasing volume growth, improving stem form, and enhancing resistance to pests and diseases. However, the significance of the wood quality and end-use potential of the material cannot be overlooked. Wood properties such as density and shrinkage due to moisture changes are crucial in the lumber and manufacturing industries. Excessive shrinkage can degrade the quality of wood products through drying defects such as warping, checking, and splitting, reducing the quality and value of wood products and potentially causing significant loss (Cáceres et al. 2018). Therefore, evaluating wood properties is critical for determining suitability for various end uses and quantifying the productive values of forest resources.

For this purpose, a range of nondestructive testing methods (NDTs) have been developed to evaluate the quality of woody materials without damaging the material, and NIR spectroscopy has gained attention as a promising tool for determining the quality of wood (Russo et al. 2019). This review discusses published studies on the use of NIR spectroscopy to evaluate wood properties. Although NIR can be applied to standing trees, research gaps and limitations remain. Some studies have noted that NIR assessments often require wood cores or sawdust samples to accurately determine wood properties.

2 Silvicultural Practices and Wood Quality

Wood is an important natural resource and a renewable material. It provides solid raw material for products such as lumber, plywood, fiber for pulp, and particles for composite products. As the deforestation of natural forests continues to raise significant economic and environmental concerns, the establishment of forest plantations offers a viable solution to ease pressure on natural forests while supplying raw materials of adequate quality for sawn timber, energy products, and wood panels. According to Russo et al. (2019), increasing the economic value of forests depends on implementing appropriate silvicultural practices in forest plantations. These practices, such as spacing, fertilization, pruning, thinning, and initial planting, can significantly influence tree growth and characteristics (Fernández et al. 2017; Kerr and Haufe 2011; Mochan et al. 2009; Raymond et al. 2008). Moreover, integrating genetic improvement with silvicultural treatments has been shown to increase fiber production per unit area and shorten rotation cycles (Alan 2020).

Eucalyptus globulus and *E. nitens* are the most widely planted species in temperate and Mediterranean zones because of their potential to produce high-quality sawn logs (Forrester et al. 2010, 2013). Other species are commonly established in subtropical and tropical zones (Forrester et al. 2010). Most plantations are intensively managed, often without thinning, and harvested as early as six years in some subtropical settings (Forrester et al. 2010). However, forest growth and wood quality can be improved through various treatments, such as thinning, pruning, and fertilizer application (Forrester et al. 2013; Potts et al. 2011; Rocha et al. 2019).

Thinning is an essential silviculture operation used to reduce tree density in a stand, improve the growth of remaining trees through sustainable approaches, and reduce long-term tree mortality rates (Schimleck et al. 2019; Tran 2014). These effects can be achieved by increasing light availability, air circulation, and temperature through canopy openings (Prasetyo et al. 2017). Following thinning, the remaining trees typically increase in diameter and develop larger crowns and branches, depending on thinning intensity and potential growth. In assessing thinning responses, long-term yield observations are considered more reliable than short-term measurements (Russo et al. 2019).

However, the increased growth rate due to thinning negatively affects end-use wood properties. A relatively high percentage of juvenile wood often makes such stands undesirable for timber processing. This issue can be avoided by the selective removal of the fastest-growing trees, a practice known as "thinning from above." This practice will eventually lead to higher basic density, lower ring width, and less taper wood (Pape 1999).

Other studies have reported more pronounced effects of thinning on wood mechanical properties. A study by Zhang et al. (2006) reported that the modulus of rupture (MOR) was lowest in trees from intensively thinned stands, whereas the trees from the control stands presented lower MORs than did those from moderately thinned stands. The modulus of elasticity (MOE) was found to decrease from the butt logs to the top logs because of larger and more frequent knots on top of the tree logs. Even though the MOR and MOE of Jack pine decreased, the stand quality (i.e., crown width, branch diameter, stem taper, and tree diameter) increased as thinning intensity increased.

Pruning is another common practice that affects a crown's photosynthetic capacity. Several studies have reported that pruned trees produce higher wood quality because they have fewer knots. In addition, the remaining branches have greater potential for diameter growth (Raymond et al. 2008). The effects of pruning on tree growth and size are influenced by the tree's physiological age, the developmental stage of its tissues, and the environmental conditions at the planting site. Pruning has also been shown to affect ring density, as it slows the formation of juvenile wood, promotes latewood formation, and increases tracheid length (Essien et al. 2016).

In terms of plant spacing practices, wider spacing influences stand leaf area and accelerates biomass development, allowing the site to be fully occupied more quickly. Tree diameter is influenced by plant spacing within a specific limit; for example, the greater the spacing area is, the less competition there is among trees and the greater the effect on the diameter of trees. Competition for sunlight shapes stem form and controls crown architecture by restricting branch sizes and accelerating branch shedding rates, thereby improving wood quality (Dobrowolska et al. 2011).

A study examining the effects of wide spacing on European ash has revealed an increased percentage of latewood formation due to increases in bole volume and mean ring width (Dobrowolska et al. 2011). This increase results in a reduction in the percentage of heartwood because the target diameters are achieved more rapidly.

The wood quality from wide-spaced European ash was reported to be denser and stronger, with improved mechanical properties such as higher modulus values. This finding aligns with Rocha et al. (2016), who reported that wider spacing in *E. grandis* × *E. camaldulensis* results in denser wood. Similarly, a study by Moulin et al. (2015) on *E. grandis* × *E. urophylla* indicated that greater spacing may increase lignin and hemicellulose contents. However, wide spacing may also slow natural pruning and increase the likelihood of wood shrinkage and swelling (Dobrowolska et al. 2011).

3 Nondestructive Testing (NDT) Application in Forestry

The process of characterizing wood materials often requires intensive qualitative and quantitative analytical methods. However, conventional methods are typically labor intensive and rely on destructive analysis of a large number of samples (Sharma et al. 2020). As the global supply of wood shifts from old-growth forests to intensively managed plantations, there is an increasing need to evaluate tree quality prior to harvest (Proto et al. 2017). With advancements in technology, various wood properties (basic density, shrinkage, compressive strength, modulus of elasticity, modulus of rupture, and microfibril angle) can now be evaluated via nondestructive methods (Proto et al. 2017; Raymond et al. 2008; Russo et al. 2019; Sharma et al. 2020; Taskhiri et al. 2020).

NDT is defined as "*the science of identifying the physical and mechanical properties of a piece of material without altering its end-use capabilities and then using this information to make decisions regarding appropriate applications*" (Ross 2015). Nondestructive testing (NDT) methods have been widely adopted in various sectors, including steel production, fiber-reinforced concrete, composite structures, and metal welding (Taskhiri et al. 2020). A range of NDT techniques, such as spectroscopic methods, acoustic technologies, and ultrasonic and thermal testing, have been developed to assess wood quality. These techniques can be applied directly to standing trees or wood products to obtain valuable information on wood quality and properties (Proto et al. 2017; Russo et al. 2019; Taskhiri et al. 2020).

Acoustic technologies have become well-established tools for material evaluation, and their use has been widely accepted for quality control (Proto et al. 2017). With the advent of portable and user-friendly time-of-flight and resonance-based devices, the use of acoustic methods in forestry has increased significantly in recent years (Proto et al. 2017; Russo et al. 2019). Numerous studies (Ayanleye 2020; Prasetyo et al. 2017;; Russo et al. 2019) have highlighted the application of stress-wave velocity (SWV) in measuring basic density (BD), microfibril angle (MFA), modulus of elasticity (MOE), and shrinkage. These findings confirm that SWV is an effective, rapid, and nondestructive technique for predicting the physico-mechanical properties of wood.

Other acoustic technologies of similar interest have proven to be beneficial for different applications. These methods include (1) ultrasonic and thermal testing to evaluate acoustic velocities and the MOE in thinned trees and (2) time-of-flight methods to detect hidden branch traces and knots (Proto et al. 2017; Taskhiri et al. 2020). Both ultrasonic and thermal testing are practical and low-cost methods for assessing the wood quality of standing trees (Proto et al. 2017). On the basis of the findings reported by Taskhiri et al. (2020), the ultrasonic systems employed in their experiments could be used to detect internal defects in standing trees in plantation settings.

In recent years, interest in near-infrared (NIR) spectroscopy methods has increased, as demonstrated in numerous studies (Ayanleye 2020; Cown 2016; Meder 2015; Sharma et al. 2020). NIR spectroscopy enables the accurate and rapid prediction of the physicochemical and chemical properties of wood through calibrated and validated measurements. Recent review findings confirm the reliability of NIR spectroscopy in predicting a wide range of wood properties.

4 Near-Infrared Spectroscopy for Determining Wood Properties

NIR is currently the most commonly used tool and is well known as an inexpensive and rapid alternative to traditional laboratory methods for estimating wood properties (Han et al. 2016; Hein 2012; Meder 2015; Potts et al. 2011; Wu et al. 2018). Conventional techniques for determining physical wood properties are costly and require destructive felling of the trees, in which NIR has advantages as a nondestructive method.

NIR spectroscopy involves various types of instruments suitable for assessing wood properties, such as Fourier transform infrared (FTIR) spectrometers, Ocean Optic systems, and X-ray diffraction (XRD). The basic procedure for acquiring empirical data prior to NIR calibration is shown in Fig. 1 (Chaix 2011; Tran 2014). Once the calibration dataset is established, future predictions can be made using the NIR alone, eliminating the need for further destructive empirical testing (Ayanleye 2020).

According to Ayanleye (2020), calibration of wood properties via NIR is typically performed via the test-set validation method. The calibration procedure for Near-Infrared Spectroscopy (NIRS) involved acquiring spectral data from wood samples using an NIR spectrometer equipped with a tungsten halogen light source, optic fiber, spectrometer, and computer system. The samples were scanned across a wavelength range of 350–2500 nm, covering both visible and near-infrared regions. For each specimen, multiple scans were collected and averaged to reduce spectral noise and obtain a representative spectrum for subsequent calibration and prediction of wood properties. Predictive models are developed on the basis of the training set, with some samples (40%) of the data randomly selected to form a validation (or test) set. These samples are then used to determine the ability of the model to predict new samples.

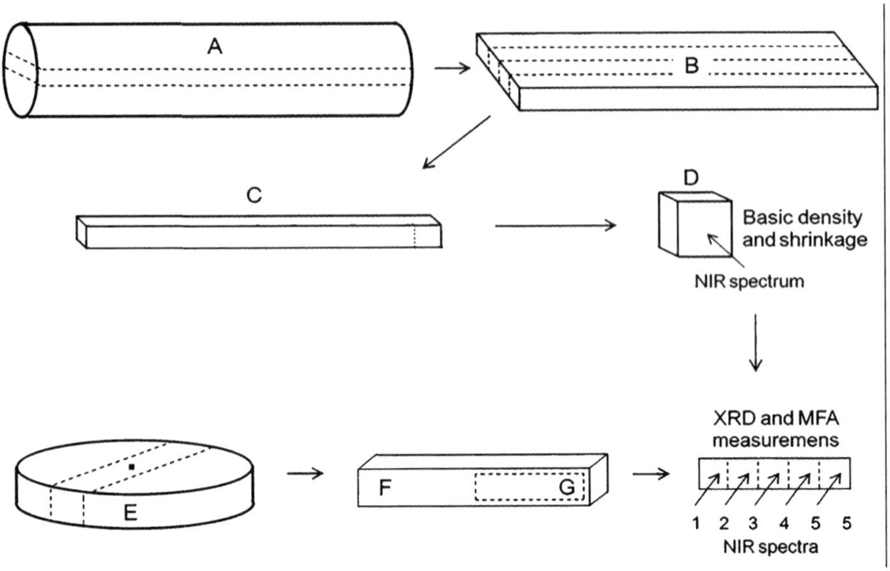

Fig. 1 Sample preparation for the measurement of basic density, radial and tangential shrinkage via the application of X-ray diffraction patterns and NIR spectra (Hein 2012)

The study also concluded that NIR spectroscopy coupled with partial least squares regression (PLSR) is a suitable technique for estimating wood properties. Figure 2 shows the regression coefficient produced from the study to predict the radial and tangential drying of *Eucalyptus* hybrid wood.

The advancement of portable NIR equipment has opened new possibilities for wood quality assessments. For example, Pederson et al. (2014) explored the application of a MicroNIR spectrometer to identify narcotic substances. The device was able to detect individual compounds from an unknown mixture via a spectral library. The approach was considered novel in pursuit of a modern apparatus, which is accessible, mobile, affordable, and easy to use, offering promising applications for in situ wood quality measurement.

NIR spectroscopy has been widely used to estimate wood properties quickly, simply, and at low cost. Baettig et al. (2017) demonstrated that a fiber optic NIR spectrometer paired with a high-precision displacement system can produce highly accurate measurements. They achieved high-level precision measurements of intra ring density values of *Pinus radiata* via the NIR/PLSR method.

Various investigations on predicting *Eucalyptus* wood density via NIR are summarized in Table 1. These studies are listed according to the values of basic density derived via the conventional method (water displacement method), alongside the coefficients of calibration (R^2 values) obtained via NIR spectroscopy.

An earlier study by Schimleck et al. (1999) reported relatively high prediction errors for *E. globulus*, with underestimation and overestimation values derived from NIR readings corresponding to low and high values of basic density, respectively. In

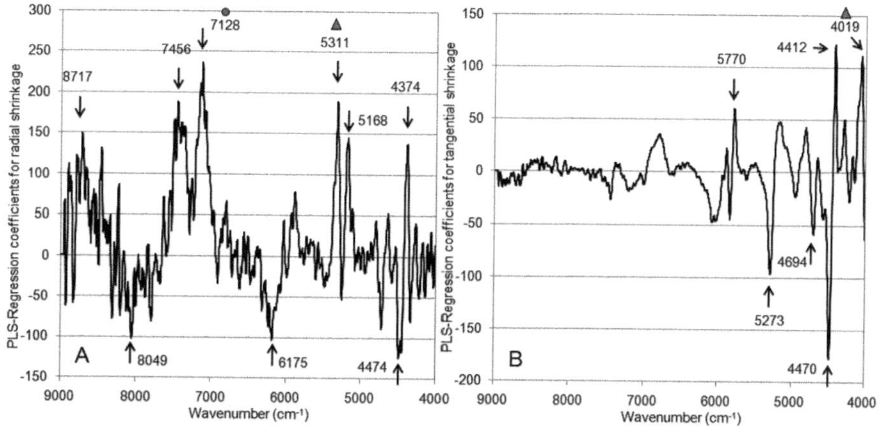

Fig. 2 Two panel figure showing PLS Regression coefficient plotted against wavenumber (cm-1). Panel A displays coefficient for radial shrinkage, with notable peaks and troughs labeled at specific wavenumbers such 7128 and 5311. Panel B shows coefficient for tangential shrinkage, with keys points marked at 4019. Both panels highlight significant data points with arrows and symbols, illustrating variations in regression coefficient across the spectrum. Bands corresponding to lignin, cellulose and hemicellulose are denoted by circles, triangles and squares respectively. (Hein 2012)

Table 1 Basic density and the coefficient of determination (R^2) from certain wavelengths of NIRS

Species	Wavelength range (nm)	Basic density, kg/m³ (12%)	R^2 value	Author
E. urophylla	800–2500	546	0.73–0.86	Hein et al. (2009)
E. delegatensis	1100–2500	–	0.77	Schimleck et al. (2001)
E. globulus	1900–2500	378–656	0.38–0.59	Schimleck et al. (1999)

a subsequent study by the same authors in 2001, they achieved much greater accuracy via the same technique for *E. delegatensis*, with R^2 values of 0.90 for MOE and 0.93 for density (Fig. 3a, b). Similarly, Hein et al. (2009) demonstrated that *the* basic density of E. urophylla could be predicted with high accuracy and efficiency via NIR. The study emphasized the importance of sample preparation and aligning fiber orientations to prevent spectral distortion due to fiber tilt. The authors also advise the use of spectral filters, the removal of anomalous samples, and the optimization of wavelength selection to improve the calibration accuracy, reduce the standard errors of calibration and cross-validation (SEC and SECV), and increase R^2 values.

Kothiyal and Raturi (2011) investigated NIR spectroscopy for predicting the specific gravity, MOE, and MOR of 5-year-old *E. tereticornis* wood. Strong correlations were found between the values measured in the laboratory and the NIR-predicted values, with R^2 values of 0.82, 0.83, and 0.81 for the MOE, MOR, and specific gravity, respectively.

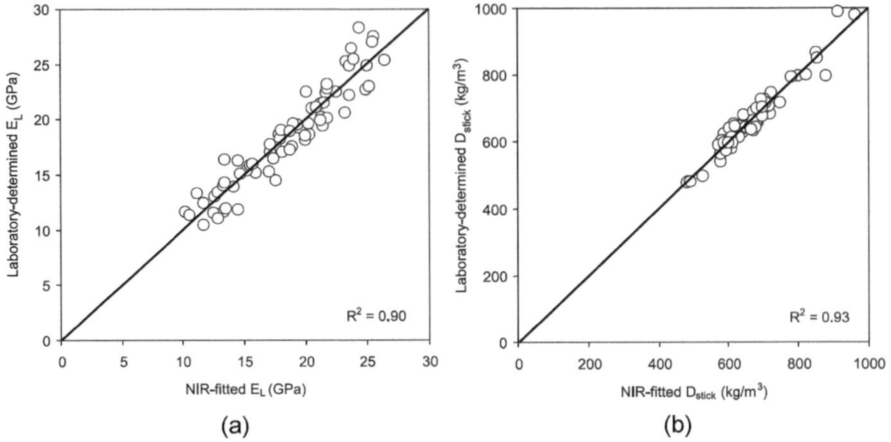

Fig. 3 Relationship between laboratory-determined (**a**) longitudinal modulus of elasticity (EL) and (**b**) density (D) versus NIR-fitted for the calibration set (Schimleck et al. 2001)

The NIR technique has also been applied for the rapid prediction of microfibril angle (MFA) variation in cores (Schimleck and Evans 2002). The study suggested that MFA calibrations be developed using samples from a range of sites and tree ages to establish a robust calibration. Inagaki et al. (2012) further demonstrated that fiber length in *E. camaldulensis* wood could be accurately predicted via the NIR/PLS method. The results met the performance thresholds specified by the AACC Method 39-00, supporting its use in screening and breeding programs.

Overall, NIR spectroscopy has proven effective and valuable for assessing multiple wood properties, including chemical, physical, mechanical, and anatomical traits. Its potential has been reviewed elsewhere (see Schwanninger et al. 2011; Tsuchikawa and Schwanninger 2013; Tsuchikawa and Kobori 2015). A growing trend of research in this area reflects a rising interest in integrating NIR spectroscopy techniques into wood science.

5 Conclusion

Extensive research has been conducted on forest plantations; however, only a few publications have explored the application of nondestructive testing (NDT) methods with wood quality assessment, particularly in relation to silvicultural practices. Recent developments in near-infrared (NIR) spectroscopy have shown promising findings, highlighting the potential for rapid and accurate analysis of chemical and mechanical wood properties. The next step in advancing the field is to expand the current knowledge base at the intersection of silviculture practices, wood quality, and NDT by validating existing findings through replication studies involving various sites, increasing sample sizes, and refining calibration procedures to increase scientific robustness. The future of NIR application in the wood industry appears promising, offering a reliable, efficient, and practical method for evaluating wood quality.

References

Alan M (2020) Silviculture and tree breeding for planted forests. Eurasian J Forest Sci 8:74–83. https://doi.org/10.3195/ejejfs.661352

Ayanleye S (2020) Assessment of some Wood properties by near infrared spectroscopy. The University of British Columbia

Baettig R, Cornejo J, Guajardo J (2017) Evaluation of intraring wood density profiles using NIRS: comparison with the X-ray method. Ann For Sci 74. https://doi.org/10.1007/s13595-016-0597-7

Cáceres CB, Hernández RE, Fortin Y (2018) Shrinkage variation in Japanese larch (*Larix kaempferi*, [Lamb.] Carr.) progenies/provenances trials in Eastern Canada. Wood Mater Sci Eng 13:97–103. https://doi.org/10.1080/17480272.2017.1327460

Chaix G (2011) Near infrared spectroscopy and eucalyptus wood properties. IUFRO working group 2.08.03, Porto Seguro, Bahia State, Brazil. Agricultural Research for Development

Cown D (2016) Review of eucalypt Wood processing issues. Scion

Dobrowolska D, Hein S, Oosterbaan A et al (2011) A review of European ash (*Fraxinus excelsior* L.): implications for silviculture. Forestry 84:133–148. https://doi.org/10.1093/forestry/cpr001

Essien C, Cheng Q, Via BK et al (2016) An acoustics operations study for Loblolly pine (*Pinus taeda*) standing saw timber with different thinning history. Bioresources 11:7512–7521

Fernández MP, Basauri J, Madariaga C et al (2017) Effects of thinning and pruning on stem and crown characteristics of radiata pine (*Pinus radiata* D. Don). iForest—Biogeosci Forestry 10:383–390. https://doi.org/10.3832/ifor2037-009

Forrester DI, Medhurst JL, Wood M et al (2010) Growth and physiological responses to silviculture for producing solid-wood products from eucalyptus plantations: an Australian perspective. For Ecol Manag 259:1819–1835. https://doi.org/10.1016/j.foreco.2009.08.029

Forrester DI, Collopy JJ, Beadle CL, Baker TG (2013) Effect of thinning, pruning and nitrogen fertilizer application on light interception and light-use efficiency in a young *Eucalyptus nitens* plantation. For Ecol Manag 288:21–30. https://doi.org/10.1016/j.foreco.2011.11.024

Han Y, Park Y, Park JH et al (2016) The shrinkage properties of red pine wood assessed by image analysis and near-infrared spectroscopy. Dry Technol 34:1613–1620. https://doi.org/10.1080/07373937.2016.1138964

Hein PRG (2012) Estimating shrinkage, microfibril angle and density of eucalyptus wood using near infrared spectroscopy. J Near Infrared Spectrosc 20:427–436. https://doi.org/10.1255/jnirs.1005

Hein PRG, Lima JT, Chaix G (2009) Robustness of models based on near infrared spectra to predict the basic density in *Eucalyptus urophylla* wood. J Near Infrared Spectrosc 17:141–150. https://doi.org/10.1255/jnirs.833

Inagaki T, Schwanninger M, Kato R et al (2012) *Eucalyptus camaldulensis* density and fibre length estimated by near-infrared spectroscopy. Wood Sci Technol 46:143–155. https://doi.org/10.1007/s00226-010-0379-6

Kerr G, Haufe J (2011) Thinning practice a Silvicultural guide. Forestry Commission

Kothiyal V, Raturi A (2011) Estimating mechanical properties and specific gravity for five-year-old *Eucalyptus tereticornis* having broad moisture content range by NIR spectroscopy. Holzforschung 65:757–762. https://doi.org/10.1515/HF.2011.055

Meder R (2015) The magnitude of tree breeding and the role of near infrared spectroscopy. NIR News 26:8–10. https://doi.org/10.1255/nirn.1521

Mochan S, Moore J, Connolly T (2009) Using acoustic tools in forestry and the wood supply chain. Forestry Commission

Moulin JC, Arantes MDC, Vidaurre GB et al (2015) Efeito do espaçamento, da idade e da irrigação nos componentes químicos da madeira de eucalipto. Revista Arvore 39:199–208. https://doi.org/10.1590/0100-67622015000100019

Pape R (1999) Influence of thinning and tree diameter class on the development of basic density and annual ring width in *Picea abies*. Scand J For Res 14:27–37. https://doi.org/10.1080/02827589908540806

Pederson CG, Friedrich DM, Hsiung C et al (2014) Pocket-size near-infrared spectrometer for narcotic materials identification. In: next-generation spectroscopic technologies VII. International Society for Optics and Photonics. SPIE, p 91010O. https://doi.org/10.1117/12.2050019

Potts B, Hamilton M, Blackburn D (2011) Developing a eucalypt resource: learning from Australia and elsewhere. In: Workshop proceedings. Wood Technology Research Centre

Prasetyo A, Aiso H, Ishiguri F et al (2017) Variations on growth characteristics and wood properties of three *Eucalyptus* species planted for pulpwood in Indonesia. Tropics 26:59–69. https://doi.org/10.3759/tropics.ms16-15

Proto AR, Macrí G, Bernardini V et al (2017) Acoustic evaluation of wood quality with a non-destructive method in standing trees: a first survey in Italy. iForest 10:700–706. https://doi.org/10.3832/ifor2065-010

Raymond CA, Joe B, Anderson DW, Watt DJ (2008) Effect of thinning on relationships between three measures of wood stiffness in *Pinus radiata*: standing trees vs. logs vs. short clear specimens. Can J For Res 38:2870–2879. https://doi.org/10.1139/X08-124

Rocha M, Vital B, De Carneiro A et al (2016) Effects of plant spacing on the physical, chemical and energy properties of *Eucalyptus* wood and bark. J Trop For Sci 28:243–248

Rocha MFV, Veiga TRLA, Soares BCD et al (2019) Do the growing conditions of trees influence the wood properties? Floresta e Ambiente 26:e20180353. https://doi.org/10.1590/2179-8087.035318

Ross RJ (2015) Nondestructive evaluation of Wood, 2nd edn. USDA Forest Service Products Laboratory

Russo D, Marziliano PA, Macri G et al (2019) Does thinning intensity affect wood quality? An analysis of Calabrian pine in southern Italy via a nondestructive acoustic method. Forests 10. https://doi.org/10.3390/f10040303

Schimleck LR, Evans R (2002) Estimation of microfibril angle of increment cores by near infrared spectroscopy. IAWA J 23:225–234

Schimleck LR, Michell AJ, Raymond CA, Muneri A (1999) Estimation of basic density of *Eucalyptus globulus* using near-infrared spectroscopy. Can J For Res 29:194–201. https://doi.org/10.1139/x98-204

Schimleck LR, Evans R, Ilic J (2001) Estimation of *Eucalyptus delegatensis* wood properties by near infrared spectroscopy. Can J For Res 31:1671–1675. https://doi.org/10.1139/cjfr-31-10-1671

Schimleck L, Dahlen J, Apiolaza LA et al (2019) Nondestructive evaluation techniques and what they tell us about wood property variation. Forests 10. https://doi.org/10.3390/f10090728

Schwanninger M, Rodrigues JC, Fackler K (2011) A review of band assignments in near infrared spectra of wood and wood components. J Near Infrared Spectrosc 19:287–308. https://doi.org/10.1255/jnirs.955

Sharma V, Yadav J, Kumar R et al (2020) On the rapid and nondestructive approach for wood identification using ATR-FTIR spectroscopy and chemometric methods. Vib Spectrosc 110. https://doi.org/10.1016/j.vibspec.2020.103097

Taskhiri MS, Hafezi MH, Harle R et al (2020) Ultrasonic and thermal testing to nondestructively identify internal defects in plantation eucalypts. Comput Electron Agric 173. https://doi.org/10.1016/j.compag.2020.105396

Tran DH (2014) NIR for combined selection in hardwoods for both growth and Wood properties ABSTRACT NIR for combined selection in hardwoods for both growth and wood properties. The University of Queensland

Tsuchikawa S, Kobori H (2015) A review of recent application of near infrared spectroscopy to wood science and technology. J Wood Sci 61:213–220. https://doi.org/10.1007/s10086-015-1467-x

Tsuchikawa S, Schwanninger M (2013) A review of recent near-infrared research for wood and paper (Part 2). Appl Spectrosc Rev 48:560–587. https://doi.org/10.1080/05704928.2011.621079

Wu T, Fang G, Liang L et al (2018) Analysis of mixed pulping raw materials of *Eucalyptus globulus* and *Acacia mangium* by near infrared spectroscopy technique combined with LASSO algorithm. Bioresources 13:1348–1359

Zhang SY, Chauret G, Swift DE, Duchesne I (2006) Effects of precommercial thinning on tree growth and lumber quality in a jack pine stand in New Brunswick, Canada. Can J For Res 36:945–952. https://doi.org/10.1139/X05-307

Satellite-Based Forest Fire Detection: Enhancing Community Safety and Environmental Monitoring Through Remote Sensing

Sheriza Mohd Razali, Nurul Fatin Musa, and Norfaryanti Kamaruddin

Abstract Satellite-based forest fire detection systems have significantly enhanced fire monitoring in forested areas, with numerous successful applications demonstrated worldwide. The integration of remote sensing technologies, such as ground-based light detection and ranging (LiDAR), has further improved the accuracy of fire degradation assessments. LiDAR, which uses laser light to generate precise 3D models of terrain and surface features, enables detailed analysis of forest structure and topography. Remote sensing plays a critical role throughout the entire fire management cycle—from forecasting fire events to postfire assessment. During a fire, remote sensing technologies are employed for real-time detection, while burn severity maps are generated to evaluate the extent of damage. Additionally, remote sensing is used to estimate fire emissions, predict potential soil erosion, support vegetation regeneration planning, and forecast future fire risks.

1 Introduction

Natural habitats are destroyed by fire because of human activity and prolonged dry seasons. Fire is inextricably related to human protection and has a global effect on high-risk communities. Will we discover a means to live in harmony with these forces? Various human causes, such as arson, are linked to fire incidents, which are aided by climatic conditions. This chapter offers a timely overview of the global availability of data from natural satellite services for forest fire monitoring and management, coinciding with the introduction of NASA's Fire Information for Resource

S. Mohd Razali (✉) · N. Kamaruddin
Institute of Tropical Forestry and Forest Products, Universiti Putra Malaysia, Serdang, Selangor, Malaysia
e-mail: sheriza@upm.edu.my

N. F. Musa
WWF-Malaysia, 1, Jalan PJS5/28A, Petaling Jaya Commercial Center (PJCC), Petaling Jaya, Selangor, Malaysia

S. Mohd Razali et al. (eds.), *Sustainable Eucalyptus Forestry in Southeast Asia*, Forestry Sciences 88, https://doi.org/10.1007/978-3-032-21343-3_7

Management System (FIRMS), a new active fire system that can be accessed by the global community. In addition to FIRMS, numerous satellite remote sensing applications continue to serve their purpose in forest fire applications, providing data for modeling and managing forest fires, thereby enhancing community safety.

Forest fires are disasters in many countries, including Australia, Portugal, Brazil, and Spain, where they occur in *Eucalyptus* forests, as well as in parts of Southeast Asia, such as Indonesia and West Malaysia, where peatland fires are common. *Eucalyptus* is a fire-tolerant forest found mostly in southeastern Australia (Karna et al. 2019) but is also present in southwestern regions (Fernandez-Carrillo et al. 2019). In Blane, a wildfire in a Mediterranean forest in the Catalan Coastal Ridge that occurred in July 2016 burned 27 hectares of forest, displacing 400 residents and requiring the intervention of 40 ground crews and 13 aerial fighters. The fire destroyed a mixed forest of *Eucalyptus*, oak, and cork oak (Soler and Úbeda 2018).

The economic impacts of forest fires are difficult to quantify in monetary terms because they involve ecosystem services and ecological processes that are not traded on the market (Secretariat CBD 2001). However, indicative assessments can offer valuable estimations of the economic effects of forest fires. In addition to financial loss, forest fires are devastating for communities, with major social and mental health consequences. Communities face the destruction of homes, especially older buildings that lack modern fire safety codes, whereas vulnerable groups such as the elderly and disabled face disproportionate risk. Communities often grapple with postfire housing shortages and insurance challenges (Richards 2019). Furthermore, people also suffer from respiratory problems, leading to increased pressure on medical facilities in some tropical countries. Reports show that hospital admissions double during peak fire seasons compared with other times (Nepstad et al. 1999).

Aside from land-use changes, forest fires have become a significant concern in many parts of the world. According to Rachen Nolan of Western Sydney University's Hawkesbury Institute for the Environment, there should be a way to build alert systems that "get details out to people so they can start preparing their properties, make sure they do not have vegetation close to their homes, and that there is good defensible space" (Walker 2020). In this context, remote sensing technologies have increasingly played a pivotal role in this domain. In 2008, Calle and Casanova (2008) published a comprehensive overview of forest fire applications in the context of remote sensing and sensor growth, highlighting how improved techniques and the integration of new sensors into global Earth observation programs have expanded the capacity to detect the danger of fire spread, identify hotspots, determine fire thermal parameters, and map affected areas. In particular, their research extensively explored the theoretical foundations of fire detection and the practical application of the National Oceanic and Atmospheric Administration's Advanced Very High-Resolution Radiometer (NOAA-AVHRR) sensors.

Behind these scenes, the Internet of Things (IoT), which is well known for its visibility, has tremendous potential in forest fire alerts and preparedness, particularly in terms of forest fire evaluation and fighting. This is because the internet allows sensors and servers to communicate and provide real-time data on soil moisture, water levels, and other relevant environmental parameters. However, IoT approaches necessitate the installation of sensors, internet gateways, cameras, and, in some cases, towers for equipment, making them less accessible in certain contexts. In contrast, remote sensing approaches show more promise because of the availability of freely accessible data and the ability to evaluate prone areas at various resolutions. However, the most important feature is that satellites can provide near-real-time data for a wide area, such as hotspot data indicating active fires within a particular hour, which can then be combined with other satellite products to provide natural resource managers with fire preparedness information.

This chapter provides an overview of remote sensing capabilities and highlights recent satellite platforms used in multidisciplinary forest fire research. It also explores how remote sensing tools have been employed to track and evaluate forest fires in *Eucalyptus*-dominated landscapes. The chapter also includes case studies that demonstrate how remote sensing has been used to investigate historical forest fire events. Special attention is given to the impact of these applications on community protection and stakeholders' perceptions of forest fire problems in countries such as Australia and Portugal, where *Eucalyptus* forests are widespread and where wildfire events have raged. This chapter serves as a valuable resource for current and future research on forest fires in *Eucalyptus* ecosystems worldwide.

2 Remote Sensing Capabilities and Recent Satellite Technologies

The development of remote sensing began with aerial photography and has since evolved into a highly sophisticated field dominated by satellite-based observation technologies. The Advanced Very High-Resolution Radiometer (NOAA-AVHRR) and Moderate Resolution Imaging Spectroradiometer (MODIS) satellites of the National Oceanic and Atmospheric Administration have been critical in providing climatological data for climate applications. On the basis of the amount of reflected energy measured by the sensors and the amount of energy released by the sensors, remote sensing sensors are classified as either active sensors or passive sensors (Coops and Tooke 2017). Active sensors, such as light detection and ranging (LiDAR) and radar sensors, rely on self-generated energy and can be used in a wide range of environments. Aerial photography, hyperspectral imaging, and multispectral imaging, on the other hand, are sensitive to radiance reflection and rely on sunlight.

Over the years, satellite-based remote sensing has enabled numerous applications, from detecting forest damage caused by natural and human-induced events via image differencing techniques to monitoring agricultural landscapes for improved management and planning. Researchers utilize satellite data at varying resolutions to enhance vegetation indices and formulations, enabling accurate monitoring of the current conditions and vigor of green vegetation. For example, freely accessible satellite data from the MODIS and Landsat series allow for advanced applications, such as the forest water index, which is based on near-infrared (NIR) and mid-infrared (MIR) reflectance. In particular, MODIS has produced a comprehensive suite of land and ocean products that can be used to conduct scientific climate assessments at global, regional, and local scales. These products fall into three categories: radiation budget variables, environmental variables, and land cover characteristics (Xiong et al. 2006). The reliability of high-resolution imagery from satellites, such as those provided by the Système Probatoire d'Observation de la Terre (SPOT), has been demonstrated, which has been used for decades across various fields, including evergreen tropical forest and riparian studies (Zhang and Zhang 2007; Kamp et al. 2013). More recent high-resolution satellites, including IKONOS, Quickbird (with 1-meter resolution), GeoEye, and the WorldView series, have further expanded the capacity for finer-scale analysis.

MODIS continues to serve as a valuable source of freely available satellite data. Its strength lies in offering consistent coverage with a broad temporal and spatial range, which helps address database limitations in regions affected by monsoon variability and other dynamic environmental conditions (Gebremichael and Barros 2006; Clark et al. 2001; Pennec et al. 2011; Ranatunge et al. 2003). WorldView-3 (WV3) complements MODIS, which was launched in 2014 as part of the WorldView satellite array (DigitalGlobe 2014). With a spatial resolution of 1.24 m, WV3 is especially effective for identifying features in tropical forests and urban wildlife (Cho et al. 2015; Hartling et al. 2019; Li et al. 2015). WV3 has four new bands in addition to the normal RGB-NIR bands: (i) coastal (400–450 nm), (ii) yellow (585–625 nm), (iii) red-edge (705–745 nm), and (iv) near-infrared-2 (NIR2) (860–1041 nm) bands. Together, these advancements in remote sensing have contributed significantly to forest fire detection, monitoring, and response. The capabilities of these methods in forest fire detection are summarized in the flowchart below, which depicts the European Forest Fire Information System (EFFIS) fire cycle (Fig. 1).

3 Remote Sensing Applications for Forest Fire and Community Safety

3.1 Recent Fire Detection and Products

Remote sensing techniques have proven invaluable in integrating multiple climatic variables for effective fire impact assessment. The NOAA-AVHRR weather satellite provides data that support informed decision-making to save lives and property (NOAA 2021). According to the latest reports, NOAA claims to have collected 6.3 billion observations every day, with weather services issuing 1.5 million forecasts and 50,000 warnings annually (NOAA 2021).

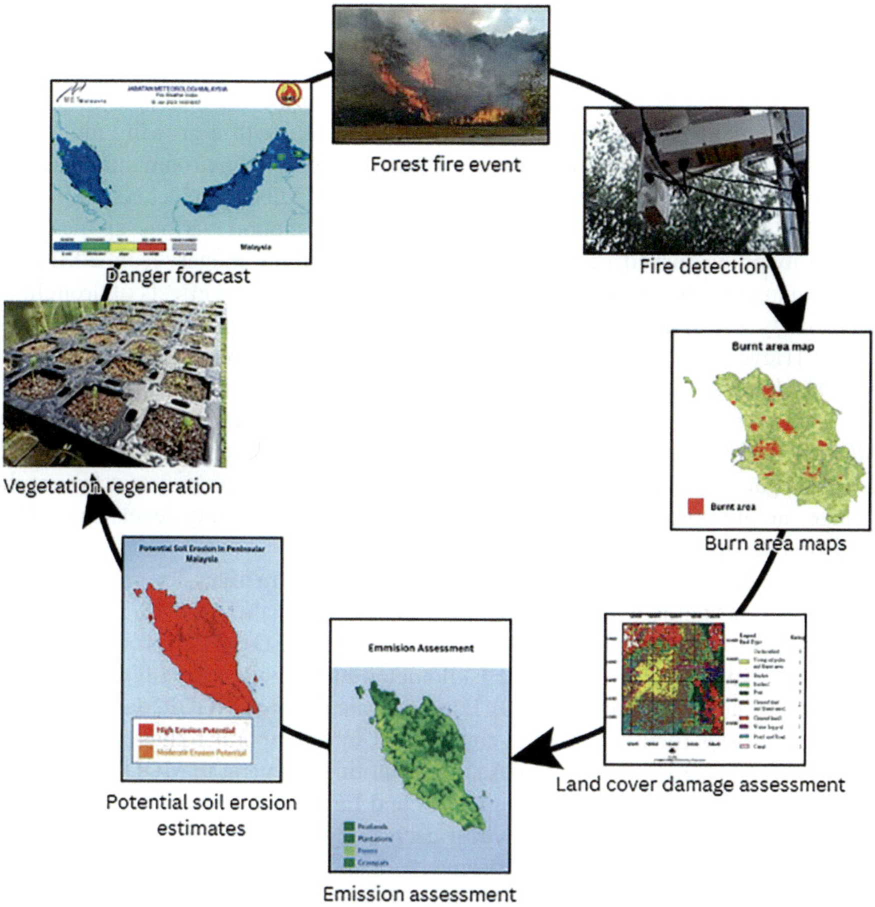

Fig. 1 Application of remote sensing in the forest fire cycle, as demonstrated in the European Forest Fire Information System (EFFIS). (Source: San-Miguel-Ayanz et al. 2012)

Indeed, NASA's Fire Information for Resource Management System (FIRMS) distributes near real-time (NRT) active fire data within three hours of satellite observation. The data are derived from NASA's Moderate Resolution Imaging Spectroradiometer (MODIS) aboard the Terra and Aqua satellites and from the Visible Infrared Imaging Radiometer Suite (VIIRS) sensors aboard the joint NASA/NOAA Suomi National Polar-orbiting Partnership (Suomi NPP) (Earthdata 2021). NASA's Earth Observing System Data and Information System (EOSDIS) established the FIRMS to address the need for rapid fire detection (Earthdata 2021), particularly for resource managers who previously faced challenges in accessing timely satellite-derived fire data. Since its launch in January 2021, the FIRMS has enabled fire managers and local communities to receive timely alerts regarding areas with significant smoke plumes, which is crucial for both emergency response and community safety (NOAA 2021).

MODIS, which was launched in 2000, incorporates a specialized algorithm for land fire detection that is based on fixed thresholds and contextual analyses. It has become a crucial sensor, succeeding NOAA-AVHRR, with a growing number of satellite products focused on enhancing human safety and environmental monitoring. One of its widely used outputs is the Vegetation Indices product NDVI Version 6 (MOD13A3) (LP DAAC 2021). These products support various applications, such as mapping burned areas, as demonstrated in MODIS imagery from 2021 tropical savanna fires (Campagnolo et al. 2021), and evaluating the effects of drought on fire activity, forest degradation, and regional smoke emissions in Bolivia's lowland forests (Heyer et al. 2018). A study by Caccamo et al. (2015) in Australia used the MOD9A1 surface reflectance product, which provides 500-meter resolution multi-day data, for fire analysis in *Eucalyptus* forests. However, the applicability of MODIS at finer spatial scales remains limited. A study conducted in Lao PDR advised against the use of MODIS active fire hotspot products to detect fires at the village, neighborhood, or plot levels, citing Tier 3 uncertainty levels that affect detection accuracy (Müller and Suess 2011).

Recent research in the United States, Australia, and Brazil has leveraged MODIS data to assess major wildfire events. The studies used the MODIS Burned Area (MCD64A1) and Global Monthly Fire Location (MCD14ML) products, which have resolutions of 500 meters and 1 kilometer, along with other MODIS datasets (Kganyago and Shikwambana 2020). Earlier applications of MODIS data in Yunnan, China, similarly focused on analyzing spatial and temporal trends in forest fire occurrence (Fornacca et al. 2017). A significant improvement in MODIS fire monitoring tools is the new Thermal Anomalies and Fire 8-Day (MOD14A2) Version 6 data, which shows the highest value of each fire pixel class found over an 8-day period. This update contains enhancements or modifications to the previous versions. More information on the MODIS products is available at https://lpdaac.usgs.gov. The thermal product was previously used in northern Sumatra to determine the pattern of hotspot distribution by village/subdistrict and land cover types, including dryland agriculture, shrubs, and bare soils adjacent to *Eucalyptus* plantation areas (Müller and Suess 2011). However, the resolution of MODIS (500 meters and 1 kilometer) limits its ability to distinguish forest types. Instead, NOAA-AVHRR and MODIS imagery and products are best suited for fire detection applications at broader, lower resolutions.

3.2 Postfire Applications in Eucalyptus *Fire-Prone Countries*

Postfire analysis is crucial for monitoring and assessing the ecological and environmental consequences of forest fires. In *Eucalyptus*-dominated, fire-prone regions, remote sensing has become increasingly indispensable for assessing fire severity, vegetation response, and ecosystem recovery. Recent studies (Braun et al. 2021;

Catry et al. 2013; Marques et al. 2011; Soler and Úbeda 2018; Volkova et al. 2017) have demonstrated the effectiveness of combining multitemporal satellite imagery to analyze fire severity. For example, freely available high-temporal-resolution Landsat imagery has been used to determine fire impacts and conduct postfire tree survival analysis in combination with synthetic aperture radar (SAR) data in the L-band. Earlier research highlighted the numerous benefits of fire risk mapping to improve fire preparedness and support better environmental and land management strategies (Oliveira et al. 2002). SAR technology, in particular, has shown tremendous potential in postfire applications because of its penetrative capabilities and weather independence, which allow detailed monitoring of burned areas (Müller and Suess 2011). This study emphasized the importance of SAR data in resolving critical issues related to forest fires, such as providing detailed hazard monitoring, thereby supporting the global community's fire mitigation efforts.

As time progresses, there is an increasing focus on evaluating fire intensity, delineating the extent of fire incidents in forested regions, and understanding the dynamics of fire danger and threat. For example, studies have examined the impact of high temperatures during wildfire events on forest foliage pigments (Soler and Úbeda 2018) and analyzed tree survival during prescribed burn events (Fernandez-Carrillo et al. 2019). These investigations have shown the invaluable role of remote sensing in producing insights into fire danger, threat levels, fire intensity, and fire detection, making it an essential tool in fire-prone ecosystems such as *Eucalyptus*-dominated landscapes.

3.2.1 Australia's Fire Disaster and Postfire Evaluation

Australia has witnessed a series of devastating wildfires that prompted extensive research into postfire recovery and evaluation methods. A study was conducted in the southern Jarrah Forest subregion of southwestern Western Australia (34° 14′ 52″ S, 115° 47′ 53″ E). The landscape is characterized by various *Eucalyptus* formations, including *Eucalyptus* low woodland, medium woodland, medium open, and tall open *Eucalyptus* forest, interspersed with patches of nonforest and softwood plantations. To evaluate the effectiveness of prescribed burns and track short-term postfire processes, such as tree survival, the study used L-band datasets from the Advanced Land Observing Satellite 2 (ALOS-2) Phased Array type L-band Synthetic Aperture Radar 2 (PALSAR-2) (Fernandez-Carrillo et al. 2019). Since wildfires wreaked havoc on many forest structures and surface elements, additional studies were undertaken to assess the effects of both postfire and controlled burning. One such study focused on fuel dynamics in an *Eucalyptus* open forest in Victoria, Australia, where a dry shrubby forest burned in 2013. This study assessed fuel dynamics 16–26 years postfire during the forest recovery phase (Volkova et al. 2019). These efforts to use satellite remote sensing in Australian forest fires have spurred further research into fire ecology (Fig. 2).

Airborne LiDAR is becoming useful in forest fire management and detection, as it was used to analyze crown structure recovery in a forested area 8.5 years after a

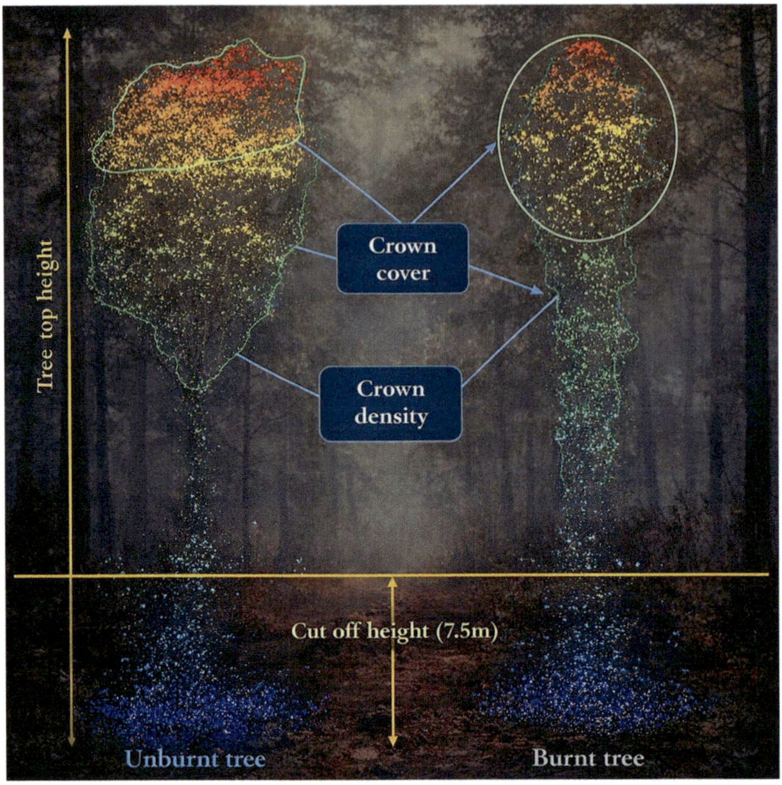

Fig. 2 Illustration of crown and crown density LiDAR metrics for individual unburnt and burnt trees derived from field measurements and LiDAR point cloud estimation (Karna et al. 2019)

2009 fire (Karna et al. 2019). The research was carried out in Victoria, approximately 100 kilometers northeast of Melbourne, Australia, using four wildfire severity categories as models in the analysis: "unburnt" (UB), "medium severity" (LS), "moderate severity" (MS), and "high severity" (HS). This study shows the effectiveness of LiDAR in detecting changes in crown structure and density following severe fire events, giving rise to recovery trajectories that are challenging to obtain with traditional optical sensors. In essence, LiDAR's high-resolution, three-dimensional data enable detailed, long-term assessment of forest dynamics. In this context, SAR and LiDAR are highly complementary tools for postfire forest analysis (Edwards et al. 2013). Furthermore, it holds promise for application in tropical forests because of its ability to capture structurally complex tree crowns, as demonstrated in dipterocarp forests across Southeast Asian countries.

3.2.2 Comparative Study of Fires in the Mediterranean Region: Portugal and Chile

The widespread forest fires that occurred in Portugal between 2005 and 2008 prompted significant postfire assessments in *Eucalyptus globulus* stands. A report released 3 years after the events analyzed damage via data from the multispectral scanner (MSS), Landsat Thematic Mapper (TM), and Landsat Enhanced Thematic Mapper Plus (ETM+). According to Marques et al. (2011), incorporating information from forest managers, such as variables related to forest density and average tree diameter, can significantly improve the predictive power of fire risk models. These variables were found to be strong indicators of postfire stand damage and tree mortality. The models developed in this study proved useful in estimating the degree of damage in a stand and predicting whether mortality will occur in the stand following a wildfire event. This report suggests the need for further data collection for improved modeling of postfire stand damage and tree mortality.

Years later, the SIMPLOT was developed as a forest simulator for *Eucalyptus* forests, which was driven primarily by wood demand. Barreiro and Tomé (2011) created a model to simulate the effects of fire intensity on forest sustainability in Portugal, aiding the Portuguese government in forecasting potential forest resource depletion. This model also simulated the effects of wildfires on the standing volume and carbon stock of *Eucalyptus* plantations.

In comparison, the study conducted in Chile is more recent, although Portuguese research remains relevant because of its emphasis on temporal imagery analysis. In Chile's *Eucalyptus* forests, researchers have studied the effects of land use change and the 2017 forest fire disaster (Braun et al. 2021). Chile is home to numerous biodiversity hotspots, with Central Chile being one of the most ecologically significant regions in Latin America. This study examined Región del Maule and Región del Biobio to determine the effects of land use change drivers on forest fire incidents. Using Landsat data from time series, they generated a plantation age map (PAM), a plantation type map (PTM), a digital elevation model (DEM), and biodiversity estimates (BES) to map fire intensity across the region. The analysis spanned land use changes from 1975 to 2015, culminating in an assessment of the ecological impact of the 2017 wildfires. The study concluded that land use change is a major driver of habitat loss and landscape fragmentation, which was further intensified by the 2017 fires. Importantly, the use of spatial data allows for the assessment of wildfire-affected areas regardless of the timing of the fire events.

Further evidence of recurring fire activity in Chile was captured through MODIS satellite imagery. A MODIS image captured a fire disaster in late January 2020, revealing thick smoke streaming over the Pacific Ocean. In early February 2017, another major fire raged across the Chilean countryside (www.earthobservatory.nasa.gov.my), and on March 29, 2015, MODIS imagery recorded a fire in southern Chile. Moreover, similar forest fire dynamics have been observed in India's deciduous dry forests in lowlands and coniferous or pine forests at relatively high elevations, which burn regularly (Kanga and Singh 2017). FARSITE (Fire Area Simulator), a model for simulating fire growth, was employed to simulate the spread and behavior of fires spatially and temporally in heterogeneous terrain, considering variables such as elevation, slope, and aspect, to mitigate forest fire damage in the region (Kanga and Singh 2017).

4 Conclusion

Given the significant economic, social, and ecological impacts of forest fires, effective management is imperative. Remote sensing has emerged as a crucial tool for monitoring and managing forest fires, particularly in *Eucalyptus* forests and other forest types. The viability and cost-effectiveness of the technology, especially for integrating local systems tailored to forest size, make it highly advantageous. Considering the recurring nature of forest fires, strengthening postfire evaluation is essential for assessing the extent of forest destruction and air pollution. Government incentives should be a new strategy in developing countries to encourage the adoption of remote sensing for forest fire management.

References

Barreiro S, Tomé M (2011) SIMPLOT: simulating the impacts of fire severity on sustainability of eucalyptus forests in Portugal. Ecol Indic 11:36–45. https://doi.org/10.1016/j.ecolind.2009.06.015

Braun AC, Faßnacht F, Valencia D, Sepulveda M (2021) Consequences of land-use change and the wildfire disaster of 2017 for the central Chilean biodiversity hotspot. Reg Environ Chang 21. https://doi.org/10.1007/s10113-021-01756-4

Caccamo G, Bradstock R, Collins L et al (2015) Using MODIS data to analyse postfire vegetation recovery in Australian eucalypt forests. J Spat Sci 60:341–352

Calle A, Casanova JL (2008) Forest fires and remote sensing. In: Coskun HG, Cigizoglu HK, Maktav MD (eds) Integration of information for environmental security. NATO Science for Peace and Security Series C: Environmental Security. Springer, Dordrecht

Campagnolo ML, Libonati R, Rodrigues JA, Pereira JMC (2021) A comprehensive characterization of MODIS daily burned area mapping accuracy across fire sizes in tropical savannas. Remote Sens Environ 252. https://doi.org/10.1016/j.rse.2020.112115

Catry FX, Moreira F, Tujeira R, Silva JS (2013) Postfire survival and regeneration of Eucalyptus globulus in forest plantations in Portugal. For Ecol Manag 310:194–203. https://doi.org/10.1016/j.foreco.2013.08.036

Cho MA, Malahlela O, Ramoelo A (2015) Assessing the utility WorldView-2 imagery for tree species mapping in South African subtropical humid forest and the conservation implications: Dukuduku forest patch as case study. Int J Appl Earth Obs Geoinf 38:349–357. https://doi.org/10.1016/j.jag.2015.01.015

Clark DA, Brown S, Kicklighter DW et al (2001) Net primary productivity in tropical forests: an evaluation and synthesis of existing field data. Ecol Appl 11:371–384

Coops NC, Tooke TR (2017) Introduction to remote sensing. In: Gergel SE, Turner MG (eds) Learning landscape ecology, 3rd edn. Springer, New York

DigitalGlobe (2014) WorldView-3. 1–2. https://earth.esa.int/eogateway/missions/worldview-3. Accessed on Nov 2025

Earthdata (2021) Fire Information for Resource Management System (FIRMS). https://earthdata.nasa.gov/earth-observation-data/near-real-time/firms

Edwards MC, Zaugg EC, Bradley JP, Bowden RD (2013) SAR and LIDAR fusion: experiments and applications. In: Radar sensor technology XVII. SPIE, p 87140U

Fernandez-Carrillo A, McCaw L, Tanase MA (2019) Estimating prescribed fire impacts and post-fire tree survival in eucalyptus forests of Western Australia with L-band SAR data. Remote Sens Environ 224:133–144. https://doi.org/10.1016/j.rse.2019.02.005

Fornacca D, Ren G, Xiao W (2017) Performance of three MODIS fire products (MCD45A1, MCD64A1, MCD14ML), and ESA Fire_CCI in a mountainous area of Northwest Yunnan, China, characterized by frequent small fires. Remote Sens 9. https://doi.org/10.3390/rs9111131

Gebremichael M, Barros AP (2006) Evaluation of MODIS Gross Primary Productivity (GPP) in tropical monsoon regions. Remote Sens Environ 100:150–166. https://doi.org/10.1016/j.rse.2005.10.009

Hartling S, Sagan V, Sidike P et al (2019) Urban tree species classification using a worldview-2/3 and liDAR data fusion approach and deep learning. Sensors (Switzerland) 19. https://doi.org/10.3390/s19061284

Heyer JP, Power MJ, Field RD, Van Marle MJE (2018) The impacts of recent drought on fire, forest loss, and regional smoke emissions in lowland Bolivia. Biogeosciences 15:4317–4331. https://doi.org/10.5194/bg-15-4317-2018

Kamp KV, Rigge M, Troelstrup NH et al (2013) Detecting channel riparian vegetation response to best-management-practices implementation in ephemeral streams with the use of spot high-resolution visible imagery. Rangel Ecol Manag 66:63–70. https://doi.org/10.2111/REM-D-11-00153.1

Kanga S, Singh SK (2017) Forest fire simulation modelling using remote sensing & GIS. Int J Adv Res Comput Sci 8

Karna YK, Penman TD, Aponte C, Bennett LT (2019) Assessing legacy effects of wildfires on the crown structure of fire-tolerant eucalypt trees using airborne LiDAR data. Remote Sens 11. https://doi.org/10.3390/rs11202433

Kganyago M, Shikwambana L (2020) Assessment of the characteristics of recent major wildfires in the USA, Australia and Brazil in 2018-2019 using multisource satellite products. Remote Sens 12. https://doi.org/10.3390/rs12111803

Li D, Ke Y, Gong H, Li X (2015) Object-based urban tree species classification using bitemporal worldview-2 and worldview-3 images. Remote Sens 7:16917–16937. https://doi.org/10.3390/rs71215861

LP DAAC (2021) MOD13A3 v006. MOD13A3 V006

Marques S, Garcia-Gonzalo J, Borges JG et al (2011) Developing postfire *Eucalyptus globulus* stand damage and tree mortality models for enhanced forest planning in Portugal. Silva Fennica 45:69–83. https://doi.org/10.14214/sf.32

Müller D, Suess S (2011) Can the MODIS active fire hotspots be used to monitor vegetation fires in the Lao PDR?

Nepstad DC, Moreira AG, Alencar AA (1999) Conservation and development of Brazil's tropical Forest regions flames in the rainforest: origins. Alternatives to Amazonian Fires, Impacts and

NOAA (2021) Weather. https://www.noaa.gov/weather

Oliveira DS, Batista AC, Soares RV, Slutter CR (2002) Fire risk mapping for pine and eucalyptus stands in Três Barras, state of Santa Catarina, Brazil. In: Viegas DX (ed) IV international conference on Forest fire research. MillPress, Rotterdam, p 244

Pennec A, Gond VR, Sabatier D (2011) Tropical forest phenology in French Guiana from MODIS time series. Remote Sens Lett 2:337–345. https://doi.org/10.1080/01431161.2010.507610

Ranatunge E, Malmgren BA, Hayashi Y et al (2003) Changes in the Southwest Monsoon mean daily rainfall intensity in Sri Lanka: relationship to the El Niño-Southern Oscillation. Palaeogeogr Palaeoclimatol Palaeoecol 197:1–14. https://doi.org/10.1016/S0031-0182(03)00383-3

Richards R (2019) After the fire: vulnerable community response and rebuild. Energy & Environment. Center for American Responses. Available at after the fire: vulnerable community response and rebuild. Energy & Environment. Center for American Responses. https://www.americanprogress.org/issues/green/reports/2019/07/25/472738/fire-vulnerable-communities-respond-rebuild/

San-Miguel-Ayanz J, Schulte E, Schmuck G, Camia A, Strobl P, Liberta G, Giovando C, Boca R, Sedano F, Kempeneers P, McInerney D, Withmore C, De Oliveira SS, Rodrigues M, Durrant T, Corti P, Oehler F, Vilar L, Amatulli G (2012) Comprehensive monitoring of wildfires in Europe: the European Forest Fire Information System (EFFIS). In: Tiefenbacher J (ed) Approaches to managing disaster—assessing hazards, emergencies and disaster impacts. InTech. https://doi.org/10.5772/28441

Secretariat of the Convention on Biological Diversity (CBD) (2001) Impacts of human-caused fires on biodiversity and ecosystem functioning, and their causes in tropical, temperate, and boreal forest biomes. Secretariat of the Convention on Biological Diversity

Soler M, Úbeda X (2018) Evaluation of fire severity via analysis of photosynthetic pigments: oak, eucalyptus and cork oak leaves in a Mediterranean forest. J Environ Manag 206:65–68. https://doi.org/10.1016/j.jenvman.2017.10.011

Volkova L, Bi H, Hilton J, Weston CJ (2017) Impact of mechanical thinning on forest carbon, fuel hazard and simulated fire behaviour in *Eucalyptus delegatensis* forest of South-Eastern Australia. For Ecol Manag 405:92–100. https://doi.org/10.1016/j.foreco.2017.09.032

Volkova L, Weiss Aparicio AG, Weston CJ (2019) Fire intensity effects on postfire fuel recovery in eucalyptus open forests of South-Eastern Australia. Sci Total Environ 670:328–336. https://doi.org/10.1016/j.scitotenv.2019.03.226

Walker T (2020) Wildfires: can we adapt to live with them? https://www.dw.com/en/climate-fires-risk-climate-change-bushfires-australia-california-extreme-weather-firefighters/a-54817927

Xiong X, Isaacman A, Barnes W (2006) MODIS level-1B products. In: Qu JJ, Gao W, Kafatos M et al (eds) Earth science satellite remote sensing, Berlin, Heidelberg

Zhang J, Zhang Y (2007) Remote sensing research issues of the national land use change program of China. ISPRS J Photogramm Remote Sens 62:461–472. https://doi.org/10.1016/j.isprsjprs.2007.07.002

Socioeconomic and Circular Economy Perspectives

Assessing the Socioeconomic Impact of *Eucalyptus* Plantation

Zaiton Samdin, Norfaryanti Kamaruddin, Norzanalia Saadun, and Nurul Shahirah Syazwani Mohd Alim

Abstract Forest plantations ensure that natural forests can be conserved from further degradation. They also complement natural forests in sustainable management, especially in ensuring the long-term availability of Malaysian forest resources to cater to present and future needs. The development of forest plantation programs has become increasingly important worldwide, considering the significant impact of forest plantations on local communities. *Eucalyptus* plantations have been proven to improve the socioeconomics of society by providing the local community with greater prosperity. The plantation also contributes to increased job opportunities both in the planting industry and in downstream industries. This chapter reviews the socioeconomic impact of *Eucalyptus* plantations, with a focus on their contribution to local income. The key aspects discussed include employment, income generation, infrastructure development, natural resource availability, sources of wood, and food security. Knowledge gaps in the socioeconomic aspects of *Eucalyptus* plantations are also addressed. Ultimately, this study aims to clarify the prevailing uncertainties, skepticism, and debates surrounding the socioeconomic implications of *Eucalyptus* plantations. The way forward involves promoting site-specific land use

Z. Samdin (✉)
School of Business and Economics, Universiti Putra Malaysia, Selangor, Malaysia
e-mail: zaisa@upm.edu.my

N. Kamaruddin
Institute of Tropical Forestry and Forest Products, Universiti Putra Malaysia, Selangor, Malaysia

N. Saadun
Institute of Tropical Forestry and Forest Products, Universiti Putra Malaysia, Selangor, Malaysia

Faculty of Forestry and Environment, Universiti Putra Malaysia, Selangor, Malaysia

N. S. S. Mohd Alim
School of Business and Economics, Universiti Putra Malaysia, Selangor, Malaysia

S. Mohd Razali et al. (eds.), *Sustainable Eucalyptus Forestry in Southeast Asia*, Forestry Sciences 88, https://doi.org/10.1007/978-3-032-21343-3_8

planning, community engagement, and sustainable management practices to balance the socioeconomic benefits of *Eucalyptus* plantations with their environmental impacts.

1 Introduction

Forest plantations have become indispensable to the global timber supply, contributing to approximately one-third of the world's industrial roundwood production (Jürgensen et al. 2014). The rapid expansion of planted forests is increasingly seen as a viable solution to meet the growing global demand for forest products and alleviate pressure on natural forests (Pirard et al. 2017). Current estimates indicate that planted forests cover approximately 294 million hectares globally, accounting for 7% of the total forest area (Food and Agriculture Organization (FAO) 2020). Among these, intensively managed plantations span approximately 131 million hectares, with Asia contributing more than half of this total (FAO 2020, 2022). Moreover, the area of other planted forests expanded by 66.8 million hectares between 1990 and 2020, with projections indicating a potential increase to around 334 million hectares by 2055 under favorable socioeconomic conditions (Korhonen et al. 2021). This highlights their growing significance in meeting the global wood demand and supporting sustainable forestry (FAO 2020).

Eucalyptus plantations have emerged as a dominant force in global afforestation efforts because of their fast growth rate and adaptability to a wide range of climatic and soil conditions (Varghese et al. 2024). However, concerns over the environmental impacts of large-scale *Eucalyptus* plantations, particularly with respect to water consumption and soil degradation, remain a key issue (Alemayehu and Melka 2022). Consequently, understanding the socioeconomic implications of *Eucalyptus* plantations is crucial for formulating policies that maximize benefits while mitigating environmental risks.

It has become increasingly evident over the past few decades that forest resources are vital for the livelihoods of people residing in or near forested areas (Olesen et al. 2022), as well as for the survival of forest-dependent communities (Ra and Sasaki 2013). Forests have historically contributed to economic development by supporting livelihoods, facilitating economic transitions, and promoting sustainable growth, thereby reiterating the significance of forest plantations (Agrawal et al. 2013). In the era of increasing prevalence of environmental pollution, afforestation has become even more critical (Landry and Chirwa 2011).

Eucalyptus comprises a genus of more than 890 species (de Oliveira et al. 2023) and is among the most planted tree species on Earth, primarily because of its rapid growth (Daba 2016). It occurs predominantly on acidic soils (Wang et al. 2023) in climates ranging from warm, moist, tropical lowlands to cool, temperate highlands (Crous et al. 2013; Turnbull 1999). Many *Eucalyptus* species can adapt to extreme geographical terrains when planted outside their natural habitat (Teketay 2000), a trait that distinguishes them from many other plant species that struggle to adjust to new environments. *Eucalyptus* species are facing increasing climate stress, with projections indicating significant shifts in their distribution due to climate change. Species in arid regions are particularly vulnerable, potentially losing up to 20% of their climate space by 2085. However, some species exhibit physiological acclimation, allowing them to adapt to rising temperatures (Drake et al. 2024).

The majority of *Eucalyptus* species, commonly known as *eucalypts*, are native to Australia (Varghese et al. 2024), with a select few naturally found in the Philippines, Papua New Guinea, Indonesia, and Timor. The reliance on natural resources for the survival of rural communities is undeniable, particularly in countries endowed with extensive forested areas. However, tropical forests are shrinking with rising populations, threatening the livelihoods of the very communities they are meant to support (Seymour and Harris 2019). Forest plantation offers a potential solution, although its expansion presents both opportunities and challenges. While such plantations contribute to economic growth and resource availability, they also alter local ecosystems and, consequently, the socioeconomic fabric of the local communities (Silva et al. 2019).

Eucalyptus plantations, in particular, have sparked considerable debate. On the one hand, they supply various products and services, such as chip wood, timber, and fuelwood (Zaizhi et al. 2002), making them crucial components of the global forestry industry. On the other hand, there are concerns regarding their environmental and social consequences (FAO 1996), especially for rural local communities (D'amato et al. 2017). While stakeholders, including users, growers, scholars, and policymakers, have raised valid objections regarding their ecological impact, the broader socioeconomic implications remain less explored. There is a pressing need to understand whether these plantations contribute positively to local livelihoods or pose economic and social challenges. Therefore, this chapter reviews the socioeconomic impact of forest plantations, with a particular focus on *Eucalyptus* and its effects on local communities.

2 Products and Services of *Eucalyptus*

The broad array of goods and services derived from *Eucalyptus* is one of the key reasons for its widespread preference over other tree species. For example, Dessie et al. (2019) revealed that cultivating *Eucalyptus* in woodlots at the farm level is a common practice among rural families in Ethiopia. These woodlots yield multiple products, including poles, split wood, and firewood bundles, which serve both household consumption and commercial purposes.

In addition to basic wood products, *Eucalyptus* offers an extensive range of economic and practical benefits, including building materials, charcoal production, fuelwood supply, pulp manufacturing (Sembiring et al. 2024), *Eucalyptus* oil extraction (Ahmad et al. 2023), and timber production (Lee et al. 2023). Its rapid growth rate makes it particularly valuable for paper production (Sembiring et al. 2021). The tree is highly valued for its versatility and ability to excel in the production of high-quality short fibers, vegetable charcoal, and various non-forest products (Food and Agriculture Organization (FAO) 2002). In countries such as Rwanda, smallholder farmers cultivate *Eucalyptus* primarily for fuelwood, fence posts, furniture, construction timber, and farm tools (Daba 2016). Its adaptability to both urban and rural environments further amplifies its utility, catering to household fuel needs while supporting affordable housing and fencing construction (Amare 2002).

However, while *Eucalyptus* is celebrated for its economic contributions, its environmental implications warrant careful consideration. Recent research highlights the contribution of *Eucalyptus* plantations to carbon sequestration and is therefore a valuable tool in climate change mitigation efforts (Brockerhoff et al. 2013; Sembiring et al. 2020). However, this ecological benefit is tempered by concerns over large-scale monoculture plantations, which have been associated with adverse effects on biodiversity and water resources (Alemayehu and Melka 2022).

Eucalyptus trees have garnered global attention from researchers and environmentalists for their rapid growth and diverse applications. The oil extracted from the leaves, fruits, buds, and bark of this plant is renowned for its antibacterial and antiseptic properties (Dixit et al. 2012; Abiri et al. 2022). It is widely used in both traditional and modern medicines for curing fever, cough, diabetes, respiratory ailments, insect bites, and muscle and joint pain (Silva et al. 2003; Kesharwani et al. 2018). In Ethiopia, for example, *Eucalyptus globulus* is traditionally used to treat the flu (Abiri et al. 2022), whereas Brazil and Australia are known to produce *Eucalyptus* oil commercially (Alemu 2016).

Table 1 summarizes the range of products and services derived from *Eucalyptus*, which include plywood, fuelwood, pulp, timber, medicine, perfumery, and honey production (Subramanian et al. 2012; Jaleta et al. 2016).

Table 1 Uses of *Eucalyptus*

Uses of *Eucalyptus*	Description
Wood	*Eucalyptus* is good for the production of plywood, lumber, building, and fencing posts (Dhakad et al. 2018) Construction materials, building materials, household and office furniture (Kaur et al. 2012) Pulp and paper production, valued for its high-quality fibers, smoothness, and printability in paper (Seng Hua et al. 2022)
Medicinal values and aromatic properties	*Eucalyptus* oil is rich in ethnomedicinal and therapeutic importance (Rassaeifar et al. 2013) *Eucalyptus* is used in alleviating many allergies such as bronchitis, congestion, sinus, and asthma (Bal and Bektaş 2012; Kesharwani et al. 2018) *Eucalyptus* oils are used in electric room diffusers and spray mists (Kesharwani et al. 2018)
Household consumption/ production	Source of fuelwood and cooking purposes (Zerga and Woldetsadik 2016) As an alternative source of energy
Ecosystem services	Mitigating climate change due to its capability in carbon sequestration (Brockerhoff et al. 2013; Bayle 2019) Windbreaking, aesthetic value, and landscaping (Weninger et al. 2021)

In addition to its commercial and medicinal benefits, *Eucalyptus* plays a pivotal role in environmental sustainability. It contributes to mitigating global warming through carbon sequestration (Brockerhoff et al. 2013; Bayle 2019), provides natural shelter and habitat for wildlife (Bennett 2016), and serves as a windbreak (Weninger et al. 2021). It also enhances landscape aesthetics and supports reforestation and land rehabilitation projects (Aida et al. 2019). The combination of ecological, economic, and therapeutic advantages positions *Eucalyptus* as a vital resource in both natural and human-managed environments.

3 Potential Socioeconomic Impact of *Eucalyptus* Plantation

Eucalyptus plantations offer various essential socioeconomic benefits, notably in terms of employment opportunities, financial security, and improved livelihoods. These plantations not only contribute to the economic empowerment of rural communities but also play a pivotal role in alleviating poverty (Alemayehu and Melka 2022) and reducing pressure on remaining native forests (Tadele et al. 2014). When integrated strategically with other agricultural activities, *Eucalyptus* plantations can significantly increase household incomes and support poverty reduction. Research has shown that *Eucalyptus* cultivation indeed offers substantial economic advantages for landowners and farmers (Liu and Li 2010).

(a) *Employment Generation*

Eucalyptus plantations offer a significant boost to employment, particularly in rural and semirural areas where alternative job opportunities are often limited. According to Bekele (2011), the majority of forestry activities occur in rural areas and require a large workforce for forestry operations, such as logging, construction, road maintenance, and general forestry management. This demand translates into employment opportunities across various sectors, including plantation establishment, traditional healing, formal employment in forestry companies, charcoal production, fuelwood collection, and the harvesting of non-wood forestry goods. Forest planting, in particular, provides local communities with stable employment opportunities (Pirard et al. 2017). In Mozambique, private forest plantations have significantly contributed to rural job creation (Bleyer et al. 2016), generating both direct and indirect roles in plantation operations and related industries (Landry and Chirwa 2011).

The establishment of *Eucalyptus* plantations has demonstrably benefited the local workforce as well as the local community (Alemu 2016), offering a vital lifeline to poor and unemployed households with the opportunity to fill vacancies created by these companies. Given that a significant portion of unskilled workers come from economically disadvantaged backgrounds, *Eucalyptus* plantation projects provide a valuable employment pathway, potentially creating thousands of jobs (Food and Agriculture Organization (FAO) 2011). For instance, in the Senan District of Ethiopia, households with *Eucalyptus* plantations had 40.2% higher total household income compared to those without such plantations (Belay et al. 2024). This increased income can lead to more spending and investment in local economies, thereby creating jobs.

Similarly, D'Amato et al. (2017) reported that industrial *Eucalyptus* plantations in China generated employment for local communities, with villagers engaging in plantation-related work, such as fertilization, weeding, and harvesting. However, concerns have been raised about the long-term sustainability of these jobs, as employment opportunities often depend on company hiring practices and the cyclical nature of plantation operations.

Nonetheless, the potential for employment generation is indisputable, especially in rural areas. The Kenana Sugar Cane Project in Sudan further illustrates this point, where *Eucalyptus* plantations have generated 300 jobs, contributing to the development of a newly established local community (Hamid and Abdalla 2015). This case demonstrates the capacity of *Eucalyptus* plantation companies to improve livelihoods by providing stable, even if potentially cyclical, employment opportunities for rural populations. In Ethiopia, *Eucalyptus* contributes up to 31% of total annual household income, making it a critical financial resource, especially during emergencies (Kerbo et al. 2024). *Eucalyptus* plantations offer higher economic returns compared to traditional agricultural practices, as evidenced in the northwestern highlands of Ethiopia, where the gross margins from *Eucalyptus* plantations are higher than those from food crops, making it a more lucrative option for farmers (Tesfaw et al. 2022).

(b) *Income and Financial Stability*

Eucalyptus plantations serve as a significant source of revenue for both smallholder farmers and commercial investors alike. There is a relationship between job creation and income, as employment opportunities contribute to increased financial stability (Landry and Chirwa 2011). Since afforestation efforts typically take place in rural areas, local communities benefit from working with the forest company, making *Eucalyptus* plantations a crucial source of income (Widianingsih et al. 2016). It reduces dependency on a single form of livelihood, thereby enhancing economic stability (Teka et al. 2025). The Food and Agriculture Organization (FAO 2013) highlights the case of the SAO Hill Forest Plantation in Tanzania, where afforestation efforts have not only increased small-scale farmer income but also contributed to poverty alleviation through forest-related business ventures.

Compared with agricultural crops, *Eucalyptus* plantations also play a crucial role in the forestry sector, increasing economic profitability (Zaizhi et al. 2002), with potentially greater financial benefits over the long term (D'amato et al. 2017). Revenue from *Eucalyptus* plantations is particularly significant for low-income households (Chanthalath et al. 2017). A study by Ango (2010) revealed that *Eucalyptus* trees are widely cultivated as cash crops, with local farmers selling standing trees to wood traders, who subsequently produce various wood products for resale in nearby towns. Additionally, *Eucalyptus* poles, which are prized for their high economic return (Tesfaw et al. 2022), have emerged as a key source of cash income for many households in plantation areas (Zerga and Woldetsadik 2016).

Eucalyptus trees are often considered a form of financial security, acting as a "living bank account" that can provide emergency funds during periods of financial distress. In this context, *Eucalyptus* serves as a financial safety net that allows low-income households to harvest and sell trees as a reliable backup source of cash (Zerga and Berta 2016; Kerbo et al. 2024). Birhanu and Kumsa (2018) reported that in Ethiopia's Central Highlands, *Eucalyptus* plantations account for 72% of the annual household cash income for impoverished families. Similarly, in Wolayita Sodo, Ethiopia, *Eucalyptus* plantations have been established as primary cash crops, generating significant income even during moisture-stress seasons (Alemu 2016). A review by Alemayehu and Melka (2022) further supports this, highlighting the significance of *Eucalyptus* plantations for smallholder farmers. Their analysis of multiple studies revealed that farmers in *Eucalyptus*-growing regions frequently reported greater revenue from these plantations, accounting for a substantial portion of their annual earnings, surpassing that from food crops.

(c) *Infrastructure Development*

Infrastructure development is one of the most significant socioeconomic impacts of forest plantations on rural communities, particularly in terms of improving access to essential resources. There are numerous reasons why forest plantation companies may invest in public services, including corporate social responsibility (CSR) and the necessity of facilitating their operations. Plantation companies motivated by CSR may invest in community infrastructure such as schools, health services, and transportation networks (Landry and Chirwa 2011). Empirical evidence supports a positive impact of plantation on infrastructure and social services in forest plantation-adjacent areas (Degnet et al. 2018; Alfred et al. 2020; Van Der Meer Simo 2020).

The *Eucalyptus* plantation company contributes financially to local infrastructure development (Pirard et al. 2017). For example, in the case of China's Guangdong Province, the expansion of *Eucalyptus* plantations has been associated with noticeable improvements in village infrastructure (Zaizhi et al. 2002). This includes the building of schools, electricity supplies, tap water systems, highways, bridges, and irrigation networks (Alemayehu and Melka 2022). However, experiences in other regions suggest that these benefits may not always translate into overall improvements in living conditions. In China's Guangxi Province, most rural communities lamented that, despite the establishment of plantations and the introduction of infrastructure, such as schools, healthcare facilities, and roads, their quality of life has remained unchanged or even deteriorated (D'amato et al. 2017). A systematic review of Southeast Asian plantation models indeed reveals a stark disconnect between actual improvement in livelihood outcomes and infrastructure gains. In some cases, communities report negligible improvements or even adverse effects on quality of life despite nearby infrastructure development (Van Der Meer Simo 2020).

(d) *Food Security Considerations*

The impact of *Eucalyptus* plantations on food security is complex and context-dependent. Manlosa et al. (2019) emphasized that integrating food crops with cash crops can improve food security. Traditional food crops harvested from plantations remain essential for sustaining nutrition, especially among local communities in rural areas (Muthoni and Nyamongo 2010). Cash crops, meanwhile, contribute to food security by generating higher agricultural revenues from the selling of products, thereby increasing household purchasing power and improving access to food (Salazar et al. 2015).

However, according to Kassie (2016), food security remains vulnerable to environmental pressures such as land erosion, population growth, and drought. While planting *Eucalyptus* trees on lands that are suitable for food crops could allow farmers to increase household incomes, it also presents trade-offs that must be carefully managed. The income generated from *Eucalyptus* plantations helps low-income households meet basic needs, including food and other essential expenses (Mekonnen et al. 2007), which is crucial in maintaining food security at the community level and mitigating food shortages (Daba 2016). Alemayehu and Melka (2022) further demonstrated that in Ethiopia's highlands, *Eucalyptus* plantations serve as an economic safety net, helping rural households sustain food availability and reduce their vulnerability to shortages.

In another study by Tadesse and Tafere (2017), 92% of respondents in Gudo Beret Kebele of Basona Worena district, Ethiopia, reported that the sales of wood products from *Eucalyptus* plantations provided significant economic profits. The revenue from these sales then helps smallholder farmers bridge food shortages at the household level (Kebebew 2010; Kerbo et al. 2024). These findings suggest that discouraging Eucalyptus plantations could jeopardize food security for smallholder farmers. Policymakers must, therefore, strike a balance between the economic benefits of *Eucalyptus* cultivation and food security priorities. For instance, providing the farmers with guidance on sustainable *Eucalyptus* management practices can help mitigate negative ecological impacts and enhance the positive economic benefits (Alemayehu and Melka 2022; Tesfaw et al. 2022).

(e) *Natural Resources Management and Sustainability*

The availability of natural resources, such as water and land, is a crucial element of the forestry ecosystem. Access to clean water is vital for ensuring healthy living conditions and sustaining ecosystems, especially in rural areas (Kattumuri 2018).

When integrated into agricultural systems, certain tree species can increase water use efficiency. However, fast-growing species such as *Eucalyptus* may worsen water scarcity (Daba 2016). Tadesse and Tafere (2017) reported that *Eucalyptus* plantations contribute to groundwater depletion due to the rapid growth of the tree species. The deep, extensive root system has a high capacity for water absorption, and excessive water uptake can lead to reduced water availability and even drying out of accessible water reserves across watersheds (Chanie et al. 2013). A study by Zegeye (2010), which assessed *Eucalyptus* plantation implications in Ethiopia, reported that if planting is not well planned, it can significantly reduce groundwater levels and adversely affect local water resources. The review by Alemayehu and Melka (2022) further highlights that *Eucalyptus* plantations caused a decline in streamflow, particularly in the highlands of Ethiopia, with the impact being more pronounced during the dry season. Daba (2016) also reported that *Eucalyptus* plantations can cause nearby springs to dry, disrupt water flow rates, and negatively impact crop production.

When planted on or near farmland, *Eucalyptus* trees can suppress crop growth and reduce yields (Chanie et al. 2013). Their expansion also affects livestock production by encroaching on grazing areas. Jaleta et al. (2016) argued that the reduction in grassland due to *Eucalyptus* expansion can lead to overgrazing, which accelerates land degradation. Over time, the contraction of natural vegetation and the expansion of barren land further contributed to environmental degradation, biodiversity loss, and disruptions in the natural ecosystem. In the case of Gudo Beret Kebele, however, where *Eucalyptus* species are mostly planted by local small-scale farmers, substantial parts of their agricultural lands have been converted into *Eucalyptus* woodlots (Tadesse and Tafere 2017). This shift is driven by several factors, including the fast-growing nature of *Eucalyptus*, declining soil productivity, and the high demand for fuelwood, construction materials, and furniture products, such as windbreaks and fencing, making *Eucalyptus* cultivation an attractive option for farmers despite their environmental drawbacks (Bekele 2015).

Therefore, to balance economic benefits with environmental sustainability, it is essential to adopt sustainable forestry practices. This includes proper land use planning, species selection, and management techniques that consider local ecological conditions. Developing and implementing policies that integrate economic growth with environmental conservation can optimize the benefits of *Eucalyptus* plantations. This includes providing support for sustainable practices and ensuring that plantations are established in suitable areas (Alemayehu and Melka 2022; Sembiring et al. 2025). As the literature reviewed here demonstrates, *Eucalyptus* plantations can deliver significant economic and environmental benefits, provided they are managed sustainably. With appropriate management practices and policy frameworks in place, the positive impacts of *Eucalyptus* on natural resources management and sustainability can be maximized, contributing to both local livelihoods and global environmental goals.

4 Conclusion

Eucalyptus plantations provide substantial socioeconomic development by generating employment, ensuring income stability, and supporting infrastructure development. In many ways, socioeconomic impacts parallel those of general afforestation efforts. These impacts also include the availability of natural resources and improvements in food security. While the overall effects are largely positive, certain aspects of *Eucalyptus* plantations may pose challenges to local communities.

Over the past century, *Eucalyptus* has expanded rapidly, becoming the most widely planted tree species worldwide. Its widespread adoption has been driven by strong support from foresters, landowners, growers, and the wood industry due to its significant economic benefits. However, a balanced approach that integrates economic viability with sustainable environmental practices is essential. Research has demonstrated that *Eucalyptus* plantations serve as a financial safety net for many impoverished farmers, particularly in areas where traditional agriculture is less viable. Future research should explore innovative plantation models that maximize socioeconomic benefits while minimizing ecological risk. Additionally, well-designed policy frameworks are necessary to ensure that *Eucalyptus* cultivation aligns with broader sustainability objectives and food security goals.

Acknowledgments This chapter forms part of the research carried out under the University Driven Research Program (UDRP) titled "Sustainable Management and Utilization of Tropical Plant Resources." The author extends sincere appreciation to all collaborators, research assistants, and local community members whose time, expertise, and support were invaluable to the successful completion of this chapter.

References

Abiri R, Atabaki N, Sanusi R, Malik S, Abiri R, Safa P et al (2022) New insights into the biological properties of eucalyptus-derived essential oil: a promising green anti-cancer drug. Food Rev Intl 38(sup1):598–633

Agrawal A, Cashore B, Hardin R, Shepherd G, Benson C, Miller D (2013) Economic contributions of forests. Backgr Pap 1:1–127

Ahmad RS, Imran M, Ahmad MH, Khan MK, Yasmin A, Saima H et al (2023) Eucalyptus essential oils. In: Essential oils. Academic Press, London, United Kingdom, pp 217–239. https://doi.org/10.1016/B978-0-323-91740-7.00005-0

Aida AAR, Arifin A, Norulhuda A, Keeren SR, Daljit SK (2019) Impact of forest rehabilitation using Eucalyptus spp. on soil properties at Sg. Betis Forest Reserve, Lojing, Kelantan. J Environ Sci Pollut Res 5(4):387–390

Alemayehu A, Melka Y (2022) Small scale eucalyptus cultivation and its socioeconomic impacts in Ethiopia: a review of practices and conditions. Trees For People 8:100269

Alemu MM (2016) Eucalyptus tree production in Wolayita Sodo, Southern Ethiopia. Open Access Libr J 3(12):1–10

Alfred K, Zaiton S, Norzanalia SA (2020) Review on the potential socioeconomic impact of eucalyptus plantation on local community. Malays For 83(2):322–339

Amare G (2002) Eucalyptus farming in Ethiopia: the case of Eucalyptus woodlots in the Amhara Region. In: Natural resources degradation and environmental concerns in the Amhara National Regional State, Ethiopia. Impact on Food Security, Proceeding, 2003, Bahar Dar, Ethiopia

Ango TG (2010) Expansion of eucalyptus plantations by smallholder farmers amid natural forest depletion: case study from Mulo District in Central Oromia. Pages 335–350 in Gil L, Tadesse W, Tolosana E, Lopez R, editors. In: Proceeding of the conference on Eucalyptus species, management, history, status and trends in Ethiopia 15–17 September 2010. Ethiopian Institute of Agricultural Research, Addis Ababa, Ethiopia (pp. 15–17)

Bal BC, Bektaş İ (2012) The effects of wood species, load direction, and adhesives on bending properties of laminated veneer lumber. Bioresources 7(3):3104

Bayle G (2019) Ecological and social impacts of eucalyptus tree plantation on the environment. J Biodivers Conserv Bioresour Manag 5(1):93–104
Bekele M (2011) Forest plantations and woodlots in Ethiopia. Afr For Forum Work Pap Ser 1:1–51
Bekele T (2015) Integrated utilization of Eucalyptus globulus grown on the Ethiopian highlands and its Contribution to rural livelihood: a case study of Oromia, Amhara and Southern Nations nationalities and people's regional state Ethiopia. Int J Basic Appl Sci 4:80–87
Belay F, Mulugeta M, Makonnen T, Shumetie A (2024) Comparative analysis of income level in matched small farms with and without eucalyptus plantations in Senan, Ethiopia. Trees For People 18:100708
Bennett AF (2016) Eucalypts, wildlife and nature conservation: from individual trees to landscape patterns. Proc R Soc Vic 128(1):71–86
Birhanu S, Kumsa F (2018) Review on expansion of Eucalyptus, its economic value and related environmental issues in Ethiopia. Int J Res Environ Sci 4(3):41–46
Bleyer M, Kniivilä M, Horne P, Sitoe A, Falcão MP (2016) Socio-economic impacts of private land use investment on rural communities: industrial forest plantations in Niassa, Mozambique. Land Use Policy 51:281–289
Brockerhoff EG, Jactel H, Parrotta JA, Ferraz SF (2013) Role of eucalypt and other planted forests in biodiversity conservation and the provision of biodiversity-related ecosystem services. For Ecol Manag 301:43–50
Chanie T, Collick AS, Adgo E, Lehmann CJ, Steenhuis TS (2013) Eco-hydrological impacts of Eucalyptus in the semi humid Ethiopian Highlands: the Lake Tana Plain. J Hydrol Hydromech 61(1):21
Chanthalath X, Yong L, Beckline M, Inthilath S (2017) Assessing the socioecological perspectives of eucalyptus cultivation and plantation expansion in Laos. Open Access Libr J 4(12):1–15
Crous KY, Quentin AG, Lin YS, Medlyn BE, Williams DG, Barton CV, Ellsworth DS (2013) Photosynthesis of temperate Eucalyptus globulus trees outside their native range has limited adjustment to elevated CO2 and climate warming. Glob Chang Biol 19(12):3790–3807
D'amato D, Rekola M, Wan M, Cai D, Toppinen A (2017) Effects of industrial plantations on ecosystem services and livelihoods: perspectives of rural communities in China. Land Use Policy 63:266–278
Daba M (2016) The Eucalyptus Dilemma: the pursuit for socioeconomic benefit versus environmental impacts of Eucalyptus in Ethiopia. J Nat Sci Res 6(19):127–137
de Oliveira DA, da Silva PHM, Novaes E, Grattapaglia D (2023) Genome-wide analysis highlights genetic admixture in exotic germplasm resources of Eucalyptus and unexpected ancestral genomic composition of interspecific hybrids. PLoS One 18(8):e0289536
Degnet MB, van der Werf E, Ingram V, Wesseler J (2018) Forest plantations' investments in social services and local infrastructure: an analysis of private, FSC certified and state-owned, non-certified plantations in rural Tanzania. Land Use Policy 79:68–83
Dessie AB, Abtew AA, Koye AD (2019) Determinants of the production and commercial values of Eucalyptus woodlot products in Wogera District, Northern Ethiopia. Environ Syst Res 8(1):4
Dhakad AK, Pandey VV, Beg S, Rawat JM, Singh A (2018) Biological, medicinal and toxicological significance of Eucalyptus leaf essential oil: a review. J Sci Food Agric 98(3):833–848
Dixit A, Rohilla A, Singh V (2012) Eucalyptus globulus: a new perspective in therapeutics. Int J Pharm Chem Sci 1(4):2020–2025
Drake JE, Vårhammar A, Aspinwall MJ, Pfautsch S, Ghannoum O, Tissue DT, Tjoelker MG (2024) Pushing the envelope: do narrowly and widely distributed Eucalyptus species differ in response to climate warming? New Phytol 243(1):82–97
Food and Agriculture Organization (FAO) (1996) The Great Eucalyptus Debate: what is it truly all about. In: Reports submitted to the regional expert consultation on Eucalyptus, vol II. Accessed from http://www.fao.org/3/AC772E/ac772e08.htm on 18 Nov 2020
Food and Agriculture Organization (FAO) (2002) Annotated bibliography on environmental, social and economic impacts of Eucalypts. Working Paper FP/17E, Rome, Italy

Food and Agriculture Organization (FAO) (2011) Eucalyptus in East Africa: socioeconomic and environmental issues, by Gessesse Dessie, Teklu Erkossa. Accessed from http://www.fao.org/3/a-am332e.pdf on 18 Nov 2020
Food and Agriculture Organization (FAO) (2013) Agribusiness public-private partnerships: a country report of the United Republic of Tanzania. Country case studies - Africa. Rome, Italy
Food and Agriculture Organization of the United Nations (2020) Global forest resources assessment 2020: main report. FAO, Rome. Accessed from https://doi.org/10.4060/ca9825en
Food and Agriculture Organization of the United Nations (2022) State of the world's forests 2022. FAO, Rome. Accessed from https://doi.org/10.4060/cb9360en
Hamid IY, Abdalla IMF (2015) Benefits of Eucalyptus camaldulensis plantations to local community in Kenana Sugar Cane Project, Sudan. Int J Curr Microbiol Appl Sci 4(3):537–541
International Congress on Planted Forests (ICPF) (2013) Planted forests are a vital resource for future green economies. Summary Report of the 3rd International Congress on Planted Forests
Jaleta D, Mbilinyi B, Mahoo H, Lemenih M (2016) Eucalyptus expansion as relieving and provocative tree in Ethiopia. J Agric Ecol Res Int 6(3):1–12
Jürgensen C, Kollert W, Lebedys A (2014) Assessment of industrial roundwood production from planted forests. Planted Forests and Trees Working Papers (FAO) Eng No. FP/48/E
Kassie GW (2016) Agroforestry and land productivity: evidence from rural Ethiopia. Cogent Food Agric 2(1):1259140
Kattumuri R (2018) Sustaining natural resources in a changing environment: evidence, policy and impact. Contemp Soc Sci 13(1):1–16
Kaur S, Pal Singh H, Batish D, Kohli R (2012) Role of monoterpenes in Eucalyptus communities. Curr Bioact Compd 8(1):101–107. https://doi.org/10.2174/157340712799828232
Kebebew Z (2010) Eucalyptus in rural livelihood safety net strategy in coffee growing area: case study around Jimına, Southwestern Ethiopia. Res J For 4(4):202–207
Kerbo AA, Degaga DT, Beyene AD (2024) Analysis of food security among eucalyptus tree farming smallholder farmers in Gurage Zone, Ethiopia: an application of a composite food security indicator. Heliyon 10(7):e29072
Kesharwani V, Gupta S, Kushwaha N, Kesharwani R, Patel DK (2018) A review on therapeutics application of eucalyptus oil. Int J Herb Med 6(6):110–115
Korhonen J, Nepal P, Prestemon JP, Cubbage FW (2021) Projecting global and regional outlooks for planted forests under the shared socio-economic pathways. New For 52(2):197–216. https://doi.org/10.1007/s11056-020-09789-z
Landry J, Chirwa PW (2011) Analysis of the potential socio-economic impact of establishing plantation forestry on rural communities in Sanga district, Niassa province, Mozambique. Land Use Policy 28(3):542–551
Lee SH, Lum WC, Antov P, Krišťák Ľ, Lubis MAR, Fatriasari W (eds) (2023) Eucalyptus: engineered wood products and other applications. Springer Nature, Singapore
Liu H, Li J (2010) The study of ecological problems of Eucalyptus plantation and sustainable development in Moaming Xiaoliang. J Sustain Dev 3:197
Manlosa AO, Hanspach J, Schultner J, Dorresteijn I, Fischer J (2019) Livelihood strategies, capital assets, and food security in rural Southwest Ethiopia. Food Secur 11(1):167–181
Mekonnen Z, Kassa H, Lemenh M, Campbell B (2007) The role and management of eucalyptus in Lode Hetosa district, Central Ethiopia. For Trees Livelihoods 17(4):309–323
Muthoni J, Nyamongo DO (2010) Traditional food crops and their role in food and nutritional security in Kenya. J Agric Food Inf 11(1):36–50
Olesen RS, Hall CM, Rasmussen LV (2022) Forests support people's food and nutrition security through multiple pathways in low-and middle-income countries. One Earth 5(12):1342–1353
Pirard R, Petit H, Baral H (2017) Local impacts of industrial tree plantations: an empirical analysis in Indonesia across plantation types. Land Use Policy 60:242–253
Ra K, Sasaki NOPHEA (2013) Assessment of local livelihood of forest-dependent communities in Cambodia. Int J Environ Rural Dev 41:63–68

Rassaeifar M, Hosseini N, Asl NHH, Zandi P, Aghdam AM (2013) Allelopathic effect of Eucalyptus globulus' essential oil on seed germination and seedling establishment of Amaranthus blitoides and Cyndon dactylon. Trakia J Sci 11(1):73–81
Salazar L, Aramburu J, González-Flores M, Winters P (2015) Food security and productivity: impacts of technology adoption in small subsistence farmers in Bolivia (No. IDB-WP-567). IDB working paper series
Sembiring N, Napitupulu HL, Sipahutar AI, Sembiring MT (2020) A review of sustainable replanting eucalyptus: higher sustainable productivity. IOP Conf Ser: Mater Sci Eng 935(1):012068. IOP Publishing
Sembiring N, Napitupulu HL, Sembiring MT, Ishak A, Gunawan HA (2021) Fulfilling Eucalyptus raw materials for pulp and paper production plants. IOP Conf Ser: Earth Environ Sci 912(1):012008. IOP Publishing
Sembiring N, Herlim C, Febrilliandika B (2024) Replanting Eucalyptus as a sustainable solution for pulp industry: a review. IOP Conf Ser: Earth Environ Sci 1352(1):012031. IOP Publishing
Sembiring N, Herlim C, Lubis AS, Sitohang DC (2025) Sustainable forestry: the dual effects of eucalyptus plantations on economy and ecology. In IOP Conference Series: Earth and Environmental Science 1445(1): 012057. IOP Publishing
Seng Hua L, Wei Chen L, Antov P, Kristak L, Md Tahir P (2022) Engineering wood products from Eucalyptus spp. Adv Mater Sci Eng 2022(1):8000780
Seymour F, Harris NL (2019) Reducing tropical deforestation. Science 365(6455):756–757
Silva J, Abebe W, Sousa SM, Duarte VG, Machado MIL, Matos FJA (2003) Analgesic and anti-inflammatory effects of essential oils of Eucalyptus. J Ethnopharmacol 89(2–3):277–283
Silva LN, Freer-Smith P, Madsen P (2019) Production, restoration, mitigation: a new generation of plantations. New For 50(2):153–168
Subramanian PA, Gebrekidan A, Nigussie K (2012) Yield, contents and chemical composition variations in the essential oils of different Eucalyptus globulus trees from Tigray, Northern Ethiopia. J Pharm Biomed Sci 17(11):1–6
Tadele D, Assefa A, Teketay D (2014) Effect of Eucalyptus camaldulensis stand conversion into crop production on growth and yield of maize: the case of Koga watershed areas in northwestern Ethiopia. Momona Ethiop J Sci 6(1):58–69
Tadesse SA, Tafere SM (2017) Local people's knowledge on the adverse impacts and their attitudes towards growing Eucalyptus woodlot in Gudo Beret Kebele, Basona Worena district, Ethiopia. Ecol Process 6(1):37
Teka MA, Abtew AA, Taw TB, Mesfin D (2025) Examining the impacts of converting commercial Eucalyptus woodlots into cropland on rural households' income and income diversification in northwestern Ethiopia. Trees For People 20:100879
Teketay D (2000) Facts and experiences on eucalypts in Ethiopia and elsewhere: ground for making life informed decisions. Walia 2000(21):25–46
Tesfaw A, Alemu D, Senbeta F, Teferi E (2022) Eucalyptus succession on croplands in the highlands of Northwestern Ethiopia: economic impact analysis using farm household model. Resources 11(8):71
Turnbull JW (1999) Eucalypt plantations. New For 17(1):37–52
Van Der Meer Simo A (2020) Livelihood impacts of plantation forests on farmers in the Greater Mekong Subregion: a systematic review of plantation forest models. Forests 11(11):1162
Varghese M, Kamalakannan R, Suraj PG (2024) Eucalyptus: taxonomy, geographic distribution, domestication, breeding, ecology and economic importance. In: Economically important trees: origin, evolution, genetic diversity and ecology. Springer Nature, Singapore, pp 81–128
Wang Z, Xu Y, He C, Wang Z, Zhu W, Wang Z et al (2023) Impact of native tree species introduction on soil nutrient and bacterial community in Eucalyptus plantations. Eur J For Res 142(6):1369–1383
Weninger T, Scheper S, Lackóová L, Kitzler B, Gartner K, King NW et al (2021) Ecosystem services of tree windbreaks in rural landscapes—a systematic review. Environ Res Lett 16(10):103002

Widianingsih NN, Theilade I, Pouliot M (2016) Contribution of forest restoration to rural livelihoods and household income in Indonesia. Sustainability 8(9):835

Zaizhi Z, Loane B, Ze H, Zhongwei Z (2002) Socio-economic assessment of Eucalyptus plantations in Suixi county, southern China. J For Plan 8(2):57–65

Zegeye H (2010) Environmental and socio-economic implications of Eucalyptus in Ethiopia. Eucalyptus species management, history, status and trends in Ethiopia. Proceedings from the congress held in Addis Ababa, 15–17 Sept 2010, (Pohjonen 1989), pp 184–205

Zerga B, Berta A (2016) Preference, purpose, and pattern of Eucalyptus tree farming in Eza Wereda, Ethiopia. Int J Res Innov Earth Sci 3(2):30–38

Zerga B, Woldetsadik M (2016) Contribution of Eucalyptus tree farming for rural livelihood in Eza Wereda, Ethiopia. Palgo J Agric 3(1):111–117

Potential Application of *Eucalyptus* spp. as Bioproducts and the Circular Economy Impact on the Industries and Society

Kaiser Mahmood, Hanisah Kamilah Abd. Razak, Yus Aniza Yusof, Norhasnida Zawawi, and Shahrul Razid Sarbini

Abstract *Eucalyptus* spp. have been widely planted for environmental protection, addressing challenges such as water and wind erosion. In addition to ecological purposes, they also contribute to the timber industry. However, pruning generates large amounts of biomass that must be managed. *Eucalyptus* spp. possess considerable potential for producing valuable bioproducts. Fully utilising biomass could enhance the economic status of a society, especially for smallholders, farmers, and entrepreneurs. Besides, the use of *Eucalyptus* spp. by-products could reduce biomass waste. This chapter explores the potential bioproducts derived from *Eucalyptus* spp., focusing on the chemical composition of the plants, industrial applications, and the role they play in a circular economy system. It also examines the economic impact of *Eucalyptus* spp. plantations in the surrounding communities.

K. Mahmood
Halal Products Research Institute, Universiti Putra Malaysia, Serdang, Selangor, Malaysia

H. K. Abd. Razak (✉) · S. R. Sarbini
Halal Products Research Institute, Universiti Putra Malaysia, Serdang, Selangor, Malaysia

Department of Crop Science, Faculty of Agricultural and Forestry Sciences, Universiti Putra Malaysia, Sarawak Campus, Bintulu, Sarawak, Malaysia
e-mail: hanisahkamilah@upm.edu.my

Y. A. Yusof
Halal Products Research Institute, Universiti Putra Malaysia, Serdang, Selangor, Malaysia

Department of Process and Food Engineering, Universiti Putra Malaysia, Serdang, Selangor, Malaysia

N. Zawawi
Halal Products Research Institute, Universiti Putra Malaysia, Serdang, Selangor, Malaysia

Department of Food Science, Faculty of Food Science and Technology, Universiti Putra Malaysia, Serdang, Selangor, Malaysia

S. Mohd Razali et al. (eds.), *Sustainable Eucalyptus Forestry in Southeast Asia*, Forestry Sciences 88, https://doi.org/10.1007/978-3-032-21343-3_9

1 Introduction

Eucalyptus (*Eucalyptus* spp.) is a well-known and economically valuable evergreen tree that has been cultivated since the 1970s, especially in Indian states such as Andhra Pradesh, Kerala, Tamil Nadu, Mysore, Gujarat, Punjab, and Haryana (Surbhi et al. 2023). The tree is native to Australia and Tasmania and is widely distributed across subtropical and warm temperate regions (Dey and Mitra 2013; Kumar et al. 2024). *Eucalyptus* is also known as Blue-gum in English, and it has various local Indian names, such as Taliparna (Sanskrit), Safeda (Hindi), Harit Parn (Gujarati), and Karpuramaram in Tamil (Dixit et al. 2012; Hardel and Laxmidhar 2011). *E. globulus* is the most recognised species and is found extensively in subtropical regions of Bolivia, Brazil, Chile, China, India, Paraguay, Portugal, Spain, and Uruguay (Dezsi et al. 2015; Mulyaningsih et al. 2011). Currently, *Eucalyptus* spp. plantings have expanded to Linguaria, Italy (Malaspina et al. 2022). Table 1 presents different *Eucalyptus* species and their bioproduct applications across various industries.

Eucalyptus spp. can grow up to 40–70 m in height with a straight massive trunk of 0.6–2 m in diameter (Hardel and Laxmidhar 2011). The leaves are petiolate, alternate, glossy, or waxy green (Dixit et al. 2012). The fluffy stamens of the flower may exist in white, cream, yellow, pink, or red. Bark characteristics vary depending on the age of the plant and degree of furrowing, with a range of thickness, hardness, or colour (Vecchio et al. 2016).

Eucalyptus spp. offer a wide array of economic and ecological benefits. Ecologically, the trees are extensively planted as a natural defence against water and wind erosion, and they contribute to the regeneration of degraded forests. Economically, they support smallholders through timber production and the use of fresh and dried branches in floral arrangements. Regular pruning is necessary throughout the life cycle of the plant to maintain tree health, promote vigorous growth, and ensure high-quality timber. This maintenance process, while essential, generates a substantial amount of biomass, such as leaves, flowers, seeds, and bark, which is often overlooked as a valuable resource. Rather than being discarded, the biomass holds significant potential for conversion into value-added products such as fuel, essential oils, and paper pulp. For example, leaf biomass is used in essential oil extraction; woody materials from branches and stems can be processed into biochar, bioplastics, or cellulose-based packaging. Even bark has been studied for tannin extraction for application in the leather industry (Pinto et al. 2013). This capacity to convert agricultural waste into commercially viable inputs positions *Eucalyptus* spp. as key contributors to the circular bioeconomy, supporting zero-waste goals and reducing dependency on synthetic or fossil-derived resources.

Eucalyptus essential oils are especially noteworthy for their broad industrial applications, particularly in the fragrance and chemical industries. The extracts have found their way into the formulation of creams, deodorisers, detergents, lotions, perfumes, and soaps (Zhang et al. 2010; Surbhi et al. 2023). Certain species, such as *E. citriodora, E. staigeriana*, and *E. camaldulensis*, are commonly used in the

Table 1 Selected species of *Eucalyptus* and their bioproduct applications

Species	Field of application	Function	References
Eucalyptus globulus	Pharmaceutical (medicinal)	Antiviral Antimicrobial Antioxidant Antifungal Antidiabetic Anti-insecticidal Antihistamine	Ikawati et al. (2001) Dezsi et al. (2015) Sonker et al. (2017)
	Food	Preservative Flavouring agents	Mulyaningsih et al. (2011) Shah et al. (2016)
Eucalyptus radiata	Pharmaceutical (medicinal)	Antibacterial	Mulyaningsih et al. (2011)
		Analgesic	Ali et al. (2015)
Eucalyptus citriodora	Aromatherapy and perfumery	Antimicrobial Anti-inflammatory Antifungal Analgesic	Dey and Mitra (2013)
	Pharmaceutical (medicinal)	Antidiabetic	Dey and Mitra (2013)
Eucalyptus staigeriana	Aromatherapy and perfumery	Antioxidative	Mulyaningsih et al. (2011)
	Agriculture	Insecticide	Maciel et al. (2010)
Eucalyptus camaldulensis	Aromatherapy and perfumery	Antimicrobial Antifungal Antioxidant	Dey and Mitra (2013)
	Pharmaceutical (medicinal)	Antidiabetic	Dawoud et al. (2015)
Eucalyptus benthamii	Pharmaceutical (medicinal)	Anticancer	Vecchio et al. (2016)
Eucalyptus cladocalyx	Pharmaceutical (medicinal)	Anticancer	Vecchio et al. (2016)
Eucalyptus grandis	Pharmaceutical (medicinal)	Antiviral	Kourkoutas et al. (2015)
Eucalyptus tereticornis	Pharmaceutical (medicinal)	Analgesic .	Silva et al. (2003) .
Eucalyptus cinerea	Agriculture	Pesticides Herbicides	Malaspina et al. (2022)
	Pharmaceutical (medicinal)	Antimicrobial	Singh et al. (2012)
Eucalyptus nicholii	Agriculture	Pesticides Herbicides	Malaspina et al. (2022)
Eucalyptus gunnii	Agriculture	Herbicides	Malaspina et al. (2022)
Eucalyptus pulverulenta	Agriculture	Herbicides	Malaspina et al. (2022)

(continued)

Table 1 (continued)

Species	Field of application	Function	References
Eucalyptus cosmophylla	Pharmaceutical (medicinal)	Antimicrobial	Singh et al. (2012)
Eucalyptus perriniana	Pharmaceutical (medicinal)	Antimicrobial	Singh et al. (2012)
Eucalyptus viminalis	Pharmaceutical (medicinal)	Antimicrobial	Singh et al. (2012)
Eucalyptus smithii	Pharmaceutical (medicinal)	Analgesic	Ali et al. (2015)

production of aromatherapy oils and perfumes (Dey and Mitra 2013). Other than that, the *Eucalyptus* essential oils serve as natural flavouring agents in food and beverage products, including chewing gums, ice-creams, and confectionery. In the agricultural sector, they may also be applied in water disinfection processes (Geoff 2007). From a medicinal perspective, essential oils demonstrate a wide array of therapeutic properties, functioning as anticancer, antidiabetic, anti-inflammatory, antibacterial, antiviral/antifungal, antiseptic, analgesic, antimalarial, anti-helminthic, and wound-healing agents (Seid et al. 2014; Vecchio et al. 2016).

The versatile uses of nearly every part of *Eucalyptus* spp. have increasingly attracted the attention of researchers and entrepreneurs seeking to explore its full potential, particularly in applications relevant to daily life. This growing interest aligns well with the principles of the circular economy, where waste materials are repurposed rather than discarded. By maximising the value derived from *Eucalyptus* biomass, this approach not only minimises environmental impact but also supports local economic development. In particular, it holds promise for enhancing the livelihoods of smallholders, entrepreneurs, and rural communities.

The subsequent sections of this chapter are structured to first provide a brief overview of the chemical composition of *Eucalyptus*-derived bioproducts, followed by an exploration of their functional applications across the pharmaceutical, food and agricultural industries. This chapter then concludes with a discussion on the broader implications of these bioproducts in the circular economy framework and their potential societal impact.

2 Chemical Composition of *Eucalyptus* spp.

Approximately 20–80 different compounds, including alkaloids, flavonoids, propanoids, and tannins, have been identified in *Eucalyptus* spp. Volatile compounds such as 1,8-cineole (eucalyptol), citronellal, globulol, piperitone, α-gurjunene, ß-pinene, aromadendrene, allo-aromadendrene, and α-, ß-, and γ-terpinen-4-ol are primarily found in the leaves and shoots. The flowers, on the other hand, are rich in dextrin and sucrose (Stackpole et al. 2011). Other constituents found in the fruit include asparagine, borneol, caproic acid, citral, cysteine, eudesmol, fenchone, glutamic acid, glycine, p-menthane, myrcene, myrtenol, ornithine, α-terpineol, threonine, and verbenone (Boulekbache-Makhlouf et al. 2010). A major proportion of the essential oil in most *Eucalyptus* spp. consists of eucalyptol (6.3–93.2%), except in *E. citriodora,* which contains citronellal as the dominant compound (40–94.9%). The concentration of certain bioactive compounds varies depending on geographical areas and methods of extraction.

3 Application of *Eucalyptus* spp. Bioproducts in Pharmaceutical Industries

The abundance of bioactive compounds in *Eucalyptus* spp. extracts and essential oils enable diverse applications in the pharmaceutical industry (Fig. 1). Compounds such as 1,8-cineole, α-pinene, β-pinene, and limonene exhibit strong antimicrobial activities, while 1,8-cineole has demonstrated anti-inflammatory activity by suppressing cytokine production, making it useful in treating inflammation associated with steroid-sensitive disorders. Additionally, compounds like monoterpenoids,

Fig. 1 Medicinal applications of *Eucalyptus* spp.

hydrolysable tannins, and ellagic acid are known for their antioxidant properties, which can help mitigate oxidative stress. Several secondary metabolites, including euglobals, cladocalol, phenolglycosides, and macrocarpal-B, exhibit anticancer activity. Borneol, linalool, safrole, geranyl acetate, and 1,8-cineole have shown promise as anti-helminthic agents (Dhakad et al. 2018).

The highest antioxidant activity has been observed in *E. globulus* (Dezsi et al. 2015), with the fruit being the primary contributor, followed by the immature flowers. The leaves also yield essential oils with notable antioxidant properties. Globulusin A and eucaglobulin, two major derivatives of *E. globulus* extracts, significantly contribute to these properties. Among the isolated compounds, γ-terpinene exhibits the strongest antioxidant activity from the extracts of *Eucalyptus* spp. (Marzoug et al. 2011). Vecchio et al. (2016) reported that the antioxidative compounds derived from *Eucalyptus* spp. demonstrate activities higher than those of ascorbic acid through the radical scavenging assay of DPPH (2,2-diphenyl-1-picrylhydrazyl). Monoterpenoids present in *Eucalyptus* spp. leaves have also been reported to exhibit moderate to high antioxidant activity (Singh et al. 2012; Kumar et al. 2024).

Terpinen-4-ol, α-pinene, and γ-terpinene present in *Eucalyptus* spp. have shown promising anticancer effects. Terpinen-4-ol inhibits tumour cells by selectively causing necrosis and cell-cycle arrest in melanoma cells. It also inhibits the growth of colorectal, pancreatic, gastric, and prostate cancer cells in a dose-dependent manner (0.005–0.1%). α-Pinene, terpinen-4-ol, and γ-terpinene have also demonstrated cytotoxic activity against Jurkat (T leukaemia), J774A.1 (murine macrophage tumour), and HeLa (cervical cancer) cells. However, the concentration of bioactive compounds varies between young and mature leaves of *Eucalyptus* spp. The cytotoxic effects of young leaves show an inhibitory mechanism by altering the enzymatic activity of mitochondria, leading to preliminary injury that may cause cell death. Abnormal membrane permeability may also occur as a concurrent mechanism to the reaction (Vecchio et al. 2016). Another compound with potential anticancer activity is α-phellandrene (Surbhi et al. 2023).

Interestingly, *Eucalyptus* spp. also exhibit antidiabetic properties. Shah et al. (2016) reported that the intake of *Eucalyptus* spp. edible extracts at 62.5 g/kg in the diet or 2.5 g/L in drinking water could reduce hyperglycaemia. In a promising experiment, albino mice fed with an aqueous extract of *Eucalyptus* demonstrated a 90% increase in glucose conversion into glycogen. Essential oil derived from *E. globulus* leaves also showed antidiabetic effects through the protein glycation process that reduces blood sugar levels and inhibits the generation of reactive oxygen species (ROS), leading to reduced oxidative stress (Dey and Mitra 2013). In addition, postprandial hyperglycaemia in type-2 diabetic rat models was alleviated using ethanolic extracts of *E. camaldulensis*. In vitro investigation into the potential of the extracts showed inhibitory activity against α-amylase and β-glucosidase. Hence, the extract could be applied to type-2 diabetes patients who encounter insulin resistance and are prone to high postprandial glucose surges (Dawoud et al. 2015).

Another major finding is that *Eucalyptus* spp. extracts can also act as antimicrobial agents. For instance, *E. globulus* extract exhibits antimicrobial effects against both Gram-negative and Gram-positive bacteria, such as *Salmonella enteritidis*,

Escherichia coli, *Staphylococcus aureus*, *Enterococcus faecium*, *Listeria monocytogenes*, and *Candida albicans* (Vecchio et al. 2016). Notable compounds with antimicrobial properties include α-terpineol, phloroglucinol-sesquiterpene, and para-isopropyl phenol (Surbhi et al. 2023). The essential oils of *Eucalyptus* spp. were also reported to effectively inhibit *Haemophilus influenzae*, *H. parainfluenzae*, and *Stenotrophomonas maltophilia* at a concentration of 1.25 μL/mL (Kourkoutas et al. 2015).

In addition to antimicrobial functions, *Eucalyptus* spp. extracts can serve as antiseptics in treating respiratory tract infections such as the common cold, sore throat and chest infections caused by bronchitis or pneumonia (Vecchio et al. 2016). Compounds including anethole, borneol, cineole, linalool, geranyl acetate, and safrole possess notable anti-helminthic activity. Several studies reported that essential oils of *Eucalyptus* spp. can inhibit parasitic worms such as hookworm and Indian adult earthworms (*Pheretima posthuma*) at concentrations of 0.05, 0.1, and 0.15 mL/mL (Taur et al. 2010).

Pain relief is another important pharmaceutical application of *Eucalyptus* spp. Cineole, a key component of the leaf essential oil, penetrates cell tissues and functions as both an analgesic and antimicrobial agent, thereby accelerating recovery. For these reasons, *Eucalyptus* oil is commonly recommended for alleviating pain associated with rheumatism, injured ligaments, and stiff muscles (Ali et al. 2015).

4 Application of *Eucalyptus* spp. Bioproducts in Food Industries

Another major industry that utilises bioproducts based on *Eucalyptus* spp. is the food industry. In various parts of India and South Africa, *Eucalyptus* leaves have traditionally been consumed as medicinal tea, prepared by boiling the leaves in hot water. Such a practice has gained popularity for its perceived benefits in relieving sore throats, fever, and diabetes mellitus (Vecchio et al. 2016; Dey and Mitra 2013). Additionally, the essential oils have been employed as decontaminating agents in the food processing industry, considering *Eucalyptus*, specifically *E. globulus* Labille, is generally recognised as safe (GRAS) by the Expert Panel of the Flavour and Extract Manufacturers Association (FEMA) (Eisenbrand et al. 2021). At low concentrations, the oils are also used as flavouring agents in beverages and food products such as baked products, confectionery, and meat products, typically as low as 0.002% (Tyagi et al. 2014; Sonker et al. 2017).

According to the European Council regulations, essential oils from *Eucalyptus* spp. can be incorporated into food products at concentrations of up to 5 mg/kg, and up to 15 mg/kg in candies and confectionery (Batish et al. 2008). The volatile components in the essential oil serve a dual function by extending product shelf life through microbial inhibition. Additionally, the addition of *Eucalyptus* spp. essential oil in candies and chewing gum not only imparts a refreshing taste but also reduces microbial load in the oral cavity. When incorporated into minced meat at concentrations between 0.1% and 0.44%, the oils act as antimicrobial agents against *E. coli* and *S. aureus*, extending the shelf life by up to 7 days and enhancing product aroma (Djenane et al. 2011). Similar shelf-life extending effects have been reported in dairy and beverage products, such as cheese, yoghurt, and fruit juices (Surbhi et al. 2023).

Given their shelf-life prolonging capability, *Eucalyptus* spp. essential oils have been used extensively in food packaging systems as an antimicrobial agent. The presence of monoterpenes in the oil creates a protective environment in the packaging, which helps inhibit microbial growth and extend the shelf life of stored food products (Sharafati et al. 2016).

5 Application of *Eucalyptus* spp. Bioproducts in Agriculture Industries

Due to the increasing global demand for wood, industries often prefer fast-growing exotic species that offer economic advantages (Zerga 2015). Among these, *Eucalyptus* spp. are a highly profitable option compared to traditional agricultural crops (Lemenih 2010). Their adaptability to poor soils and resilience make them suitable for integrated farming systems, particularly in land rehabilitation projects and degraded forest recovery efforts (Gil et al. 2010; Muluneh 2011). In an agroforestry system, *Eucalyptus* spp. can protect against soil erosion, enhance carbon fixation, create tree-diverse agricultural landscapes, and contribute to sustainable land management methods. The abundance of bioactive compounds in the tree also provides natural pest control for the surrounding area. Besides, the natural flora and fauna will also be preserved (Singh et al. 2021).

Malaspina et al. (2022) performed a phytochemical analysis of *E. cinerea* and *E. nicholii*, identifying eucalyptol (1,8-cineole) as the dominant compound in both species (particularly abundant in *E. nicholii*). Other constituents identified included α-pinene and terpinene derivatives (α-terpinene, γ-terpinene, terpinolene, α-terpineol, and α-terpinyl acetate). The researcher noted a higher density of oil glands in the leaves of *E. nicholii* compared to *E. cinerea.* Factors influencing oil yield and chemical composition include the age of the tree and the physiological stage of the leaves (mature versus young) (Singh et al. 2012). Importantly, both species demonstrated phytotoxicity against weeds without significantly inhibiting the germination of most horticultural crops, except for radish (*Raphanus sativus*). These findings support the potential application of *Eucalyptus* extracts as natural herbicides and pesticides (Malaspina et al. 2022).

In another study, Batish et al. (2008) reported that *Eucalyptus* spp. essential oil could be used as an insect repellent. The study shows that it could repel insects for up to 8 h, depending on concentration. The insecticidal activity is attributed to the presence of 1,8-cineole, citronellal, p-cymene, eucamalol, limonene, α-pinene, γ-terpinene, and allo-ocimene (Su et al. 2006). Eucalyptol (1,8-cineole), in particular, can act as a carrier solvent for insecticides and pesticides and is the major compound responsible for the repellent attributes. Interestingly, certain species like

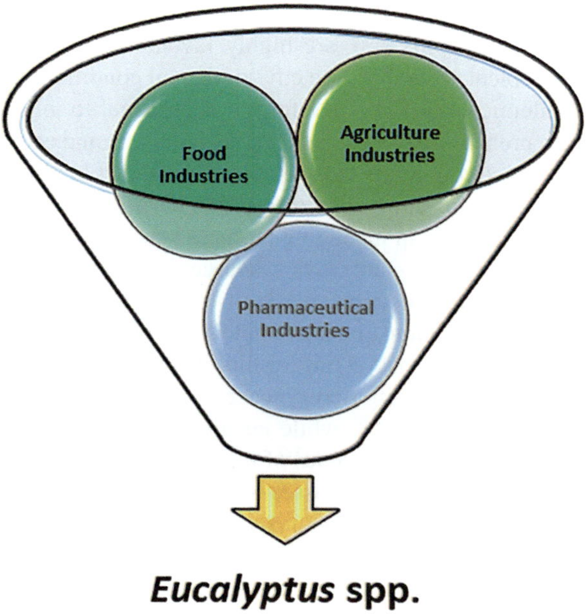

Fig. 2 Bioproducts industries of *Eucalyptus* spp.

E. staigeriana, E. citriodora, and *E. globulus* have demonstrated superior effectiveness for controlling blood-sucking sandflies (*Lutzomyia longipalpis*) compared to other species (Maciel et al. 2010). In another study, Rani et al. (2016) found that fumigation with *E. globulus* leaf oil terminated the adult *Musca domestica* houseflies within 6.5 min. The essential oil was also effective in repelling red bugs, with acaricidal effects observed at the concentration of 50% (Samani et al. 2015).

Therefore, *Eucalyptus* spp. are versatile and economically valuable trees with wide-ranging applications across various industries (Fig. 2). Their potential merits deserve further attention from policymakers, entrepreneurs, smallholders, and farmers.

6 Circular Economy Impact of *Eucalyptus* spp. Bioproducts to the Society

Eucalyptus spp. are among the most extensively cultivated tree species, commonly through government initiatives, community programmes, or private woodlots. Their broad utility, from wood to bioproducts, offers numerous benefits to society. *Eucalyptus* spp. are robust trees with fast growth, which is particularly advantageous for smallholders and entrepreneurs seeking economic returns within a relatively short cultivation period. The harvested wood is often used to make dwellings, fences, and agricultural tools, supporting basic infrastructure in rural communities in particular (Muluneh 2011). Furthermore, the diverse bioproducts derived from *Eucalyptus* spp. contribute to boosting food security, mitigating poverty, and diversifying both the incomes of the farmers and farming practices (Zerfu 2002). As such, *Eucalyptus* spp. are highly favoured by farmers, especially in tropical and subtropical regions where environmental conditions are conducive to its cultivation (Hailemicael 2012). Therefore, it is essential to inform farmers, smallholders, and entrepreneurs about the substantial economic and ecological potential of *Eucalyptus* spp., encouraging its cultivation and sustainable utilisation.

Eucalyptus cultivation also offers clear synergies with circular economy principles, especially in managing biomass waste. Two key silvicultural practices—pruning and felling—are major sources of organic biomass. Pruning is routinely performed to produce solid, knot-free wood, especially for high-value timber, and this generates large quantities of leaf and branch biomass (Forrester 2013). Similarly, industrial-scale harvesting results in substantial volumes of logging biomass. For instance, Brazil reportedly generated about 19 million cubic meters of wood and forest waste in 2018, while eucalyptus plantations of Uruguay produce between 90,000 and 140,000 tonnes of forestry biomass annually (Vallejos et al. 2017).

These residual biomasses—often discarded or underutilised—hold significant potential for conversion into value-added bioproducts such as biochar, mulching material, essential oils, or even pulp for biodegradable packaging. Such initiatives not only reduce environmental waste but also create new income avenues and industrial inputs. Hence, the circular utilisation of *Eucalyptus* spp. by-products plays a vital role in minimising ecological footprints while supporting sustainable rural economies. This approach aligns strongly with circular economy principles by transforming biomass waste into value-added products (Fig. 3), thereby minimising environmental impact and maximising resource efficiency.

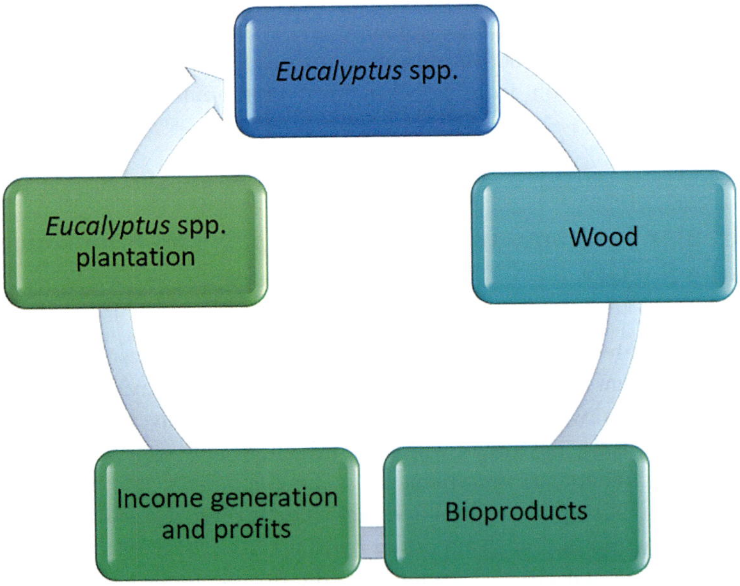

Fig. 3 The circular economy of *Eucalyptus* spp.

7 Conclusion

In summary, the cultivation of *Eucalyptus* spp. presents significant opportunities for farmers, smallholders, and entrepreneurs, particularly when implemented on a larger scale. Its fast growth, resilience, and rich phytochemical profile make it a prime resource for sustainable bioproduct development across a spectrum of sectors, from pharmaceuticals and food to agriculture and industry. The significant biomass generated through essential practices like pruning and industrial felling could be transformed into high-value products that directly support circular economy principles by maximising resource utilisation and minimising waste. Thus, *Eucalyptus* spp. play a pivotal role in promoting resource efficiency, reducing biomass waste streams, and fostering environmentally responsible economic development. Future research should explore the advancement of cost-effective biomass conversion technologies, alongside the establishment of supportive policy frameworks that incentivise the integration of *Eucalyptus* spp. in advancing global circular economy goals.

References

Ali B, Al-Wabel NA, Shams S, Ahamad A, Khan SA, Anwar F (2015) Essential oils used in aromatherapy: a systemic review. Asian Pac J Trop Biomed 5(8):601–611

Batish DR, Singh HP, Kohli RK, Kaur S (2008) Eucalyptus essential oil as natural pesticide. For Ecol Manag 256:2166–2174

Boulekbache-Makhlouf L, Meudec E, Chibane M, Mazauric JP, Slimani S, Henry M, Madani K (2010) Analysis by high-performance liquid chromatography diode array detection mass spectrometry of phenolic compounds in fruit of *Eucalyptus globulus* cultivated in Algeria. J Agric Food Chem 58(24):12615–12624

Dawoud ADH, Shayoub MEH, Shayoub SMEH (2015) Effects of ethanolic leaf extract of Eucalyptus camaldulensis on oral glucose tolerance test in Type-2 Model diabetic rats. J Netw Commun Emerg Technol 2(2):6–8

Dey B, Mitra A (2013) Chemo-profiling of eucalyptus and study of its hypoglycemic potential. World J Diabetes 4(5):170–176

Dezsi Ş, Alexandru SB, Cristina B, Dan CV, Radu SD, Ana-Maria G, Andrei M, Laurian V (2015) Antimicrobial and antioxidant activities and phenolic profile of *Eucalyptus globulus* Labill. and *Corymbia ficifolia* (F. Muell.) Hill KD, Johnson LAS. leaves. Molecules 20(3):4720–4734

Dhakad AK, Pandey VV, Beg S, Rawat JM, Singh A (2018) Biological, medicinal and toxicological significance of Eucalyptus leaf essential oil: a review. J Sci Food Agric 98(3):833–848

Dixit A, Rohilla A, Singh V (2012) *Eucalyptus globulus*: a new perspective in therapeutics. Int J Pharm Chem Sci 1(4):2020–2025

Djenane D, Yanguela J, Amrouche T, Boubrit S, Boussad N, Roncales P (2011) Chemical composition and antimicrobial effects of essential oils of *Eucalyptus globulus*, *Myrtus communis* and *Satureja hortensis* against *Escherichia coli* O157: H7 and *Staphylococcus*. Food Sci Technol Int 17(6):505–515

Eisenbrand G, Cohen SM, Fukushima S, Gooderham NJ, Guengerich FP, Hecht SS, Rietjens IM, Rosol TJ, Davidsen JM, Harman CL, Taylor SV (2021) FEMA GRAS assessment of natural flavor complexes: eucalyptus oil and other cyclic ether-containing flavoring ingredients. Food Chem Toxicol 155:112357

Forrester DI (2013) Growth responses to thinning, pruning and fertiliser application in *Eucalyptus* plantations: a review of their production ecology and interactions. For Ecol Manag 310:336–347
Geoff P (2007) Eucalyptus oil applications and markets. Science 3:1–95
Gil L, Tadesse W, Tolosana E, López R (2010) Eucalyptus species management, history, status and trends in Ethiopia. EIAR, UPM, ENCE, Madrid
Hailemicael BK (2012) The contribution of Eucalyptus woodlots to the livelihoods of small scale farmers in tropical and subtropical countries with special reference to the Ethiopian Highlands. Master Thesis, Swedish University of Agricultural Sciences, p 43
Hardel DK, Laxmidhar S (2011) A review on phytochemical and pharmacological of *Eucalyptus globulus*: a multipurpose tree. Int J Res Ayurveda Pharm 2(5):1527–1530
Ikawati, Wahyuono S, Maeyama K (2001) Screening of several Indonesian medicinal plants for their inhibitory effect on histamine release from RBL-2H3 cells. J Ethnopharmacol 75:249–256
Kourkoutas Y, Karatzas KA, Valdramidis VP, Chorianopoulos N (2015) Bioactive natural products: facts, applications, and challenges. Biomed Res Int 2015(1):684109
Kumar A, Singh V, Johar V, Durgaprasad T (2024) Eucalyptus oil as a remedy for agriculture crop protection: a review. Ecol Environ Conserv 30:S70–S73
Lemenih M (2010) Growing Eucalyptus by smallholder farmers in Ethiopia. In: Gil L, Wubalem T, Tolosana E, Lopez R (eds) Eucalyptus species management, history, status and trends in Ethiopia. Ethiopian Institute of Agricultural Research, Addis Ababa, pp 91–103
Maciel MV, Morais SM, Bevilaqua CM, Silva RA, Barros RS, Sousa RN, Sousa LC, Brito ES, Souza-Neto MA (2010) Chemical composition of *Eucalyptus* spp. essential oils and their insecticidal effects on *Lutzomyia longipalpis*. Vet Parasitol 167(1):1–7
Malaspina P, Papaianni M, Ranesi M, Polito F, Danna C, Aicardi P, Cornara L, Woo SL, De Feo V (2022) *Eucalyptus cinerea* and *E. nicholii* by-products as source of bioactive compounds for agricultural applications. Plants 11(20):2777
Marzoug HNB, Romdhane M, Lebrihi A, Mathieu F, Couderc F, Abderraba M, Khouja ML, Bouajila J (2011) *Eucalyptus oleosa* essential oils: chemical composition and antimicrobial and antioxidant activities of the oils from different plant parts (stems, leaves, flowers and fruits). Molecules 16(2):1695–1709
Muluneh M (2011) Eucalyptus plantations in the highlands of Ethiopia revisited: a comparison of soil nutrient status after the first coppicing. Master Thesis, Mountain Forestry Program, University of Natural Resources and Life Sciences, Vienna, Austria
Mulyaningsih S, Sporer F, Reichling J, Wink M (2011) Antibacterial activity of essential oils from Eucalyptus and of selected components against multidrug-resistant bacterial pathogens. Pharm Biol 49(9):893–899
Pinto PC, Sousa G, Crispim F, Silvestre AJ, Neto CP (2013) *Eucalyptus globulus* bark as source of tannin extracts for application in leather industry. ACS Sustain Chem Eng 1(8):950–955
Rani N, Harikrishnan TJ, Ponnudurai G (2016) *In vitro* insecticidal activity of essential oil of Eucalyptus globulus against Musca domestica. Indian Vet J 93(9):25–27
Samani AD, Ghahfarokhi SM, Samani AD, Kheirabadi KP (2015) Acaricidal and repellent activities of essential oil of Eucalyptus globulus against *Dermanys susgallinae* (*Acari: Mesostigmata*). J Herbmed Pharmacol 4(3):81–84
Seid Y, Manoharan J, Omprakash S (2014) Extraction of essential oil from eucalyptus leaves as antibacterial application on cotton woven fabric. IJBVI 1(1):1–7
Shah G, Bajaj J, Soni V, Dhawan RK (2016) *Eucalyptus* genus: a review. J Pharm Res 10(10):609–617
Sharafati CF, Taghizadeh M, Rafieian-kopaei M, Sharafati-chaleshtori R (2016) Effect of chitosan incorporated with cumin and eucalyptus essential oils as antimicrobial agents on fresh chicken meat. J Food Process Preserv 40(3):396–404
Silva J, Abebe W, Sousa SM, Duarte VG, Machado MIL, Matos FJA (2003) Analgesic and anti-inflammatory effects of essential oils of Eucalyptus. J Ethnopharmacol 89(2–3):277–283
Singh HP, Kaur S, Negi K, Kumari S, Saini V, Batish DR, Kohli RK (2012) Assessment of in vitro antioxidant activity of essential oil of *Eucalyptus citriodora* (lemon-scented Eucalypt; Myrtaceae) and its major constituents. LWT Food Sci Technol 48(2):237–241

Singh V, Johar V, Kumar R, Chaudhary M (2021) Socio-economic and environmental assets sustainability by agroforestry systems: a review. Int J Agric Environ Biotechnol 14(04):521–533

Sonker P, Verma S, Gupta P (2017) To study the pharmacological effect and beneficial effect of *Eucalyptus globulus* in different types of diseases. Int J Res Pharmacol Pharmacother 6(1):81–88

Stackpole DJ, Vaillancourt RE, Alves A, Rodrigues J, Potts BM (2011) Genetic variation in the chemical components of *Eucalyptus globulus* wood. G3 Genes Genom Genet 1(2):151–159

Su YC, Ho CL, Wang IC, Chang ST (2006) Antifungal activities and chemical compositions of essential oils from leaves of four Eucalyptus. Taiwan J For Sci 21:49–61

Surbhi, Kumar A, Singh S, Kumari P, Rasane P (2023) Eucalyptus: phytochemical composition, extraction methods and food and medicinal applications. Adv Tradit Med 23(2):369–380

Taur DJ, Kulkarni VB, Patil RY (2010) Chromatographic evaluation and anthelmintic activity of *Eucalyptus globulus* oil. Pharm Res 2(3):125

Tyagi A, Bukvicki D, Gottardi D, Tabanelli G, Montanari C, Malik A, Guerzoni ME (2014) Eucalyptus essential oil as a natural food preservative: *in vivo* and *in vitro* antiyeast potential. Biomed Res Int 2014(1):969143

Vallejos ME, Kruyeniski J, Area MC (2017) Second-generation bioethanol from industrial wood waste of South American species. Biofuel Res J 15:654–667

Vecchio M, Loganes C, Minto C (2016) Beneficial and healthy properties of eucalyptus plants: a great potential use. Open Agric J 10:52–57

Zerfu H (2002) Ecological impact evaluation of Eucalyptus plantation in comparison with agricultural and grazing land-use types in the highlands of Ethiopia. Ph.D. Thesis, Institute of Forest Ecology, Vienna University of Agricultural Science, Vienna

Zerga B (2015) Ecological impacts of Eucalyptus plantation in Eza Wereda, Ethiopia. Int Inv J Agric Soil Sci 3(4):47–51

Zhang J, An M, Wu H, Stanton R, Lemerle D (2010) Chemistry and bioactivity of Eucalyptus essential oils. Allelopath J 25(2):313–330

Utilization of the Bark and Leaves of *Eucalyptus urophylla* as a Source of Tannin for Natural Color Binding on Fabrics Using the Ecoprinting Technique for Supporting Circular Economy

Arif Nuryawan, Iwan Risnasari, Berry Jayusman Panjaitan, Mohd Yusoff Ishak, and Simon Sidabukke

Abstract The industrial use of *Eucalyptus urophylla* for pulp production generates significant amounts of bark and leaf residues. To support a circular economy, these residues can be utilized as a natural source of tannins for color binding in fabric dyeing via the ecoprinting technique. Compared with batik, a traditional fabric-dyeing method, ecoprinting is considered more environmentally friendly because of its minimal chemical usage, reliance on natural colorants, and absence of wax in the production process. This study aimed to compare the potential of *E. urophylla* bark and leaves as sustainable sources of tannins for biomordant applications in ecoprinting. Tannins were selected because of their strong affinity for fibers, which enhances color fastness while reducing the reliance on synthetic mordants. An example of the application of tannin, which was obtained from a commercial pulp-production company in North Sumatra, Indonesia, derived from the extraction of bark and leaves of *E. urophylla* for binding the natural dye derived from Secang wood (*Caesalpinia*

A. Nuryawan (✉) · I. Risnasari
Faculty of Forestry, Universitas Sumatera Utara, Deli Serdang, North Sumatra, Indonesia

Center of Excellence for Songket, Universitas Sumatera Utara, Medan, North Sumatra, Indonesia

Association of Eco-Printer Indonesia (AEPI), Division of Research and Development, North Sumatra, Indonesia
e-mail: arif5@usu.ac.id

B. J. Panjaitan
Faculty of Forestry, Universitas Sumatera Utara, Deli Serdang, North Sumatra, Indonesia

M. Y. Ishak
Faculty of Forestry and Environment, Universiti Putra Malaysia, Serdang, Malaysia

S. Sidabukke
PT. Toba Pulp Lestari, Sosor Ladang, Parmaksian, Toba Samosir, North Sumatra, Indonesia

Faculty of Agriculture, Universitas Simalungun, Pematang Siantar, Indonesia

S. Mohd Razali et al. (eds.), *Sustainable Eucalyptus Forestry in Southeast Asia*, Forestry Sciences 88, https://doi.org/10.1007/978-3-032-21343-3_10

sappan). Ecoprinting was performed on the fabrics via three types of leaves—*Swietenia mahagoni*, Johar (*Cassia siamea*), and Krei Payung (*Filicium decipiens*)—followed by an evaluation of color retention, leaching resistance, and color difference (ΔE) using a colorimeter. The tannin yield from *E. urophylla* bark was slightly greater than that from the leaves. Both parts contain tannins, along with other phytochemicals, such as alkaloids, flavonoids, terpenoids, steroids, and saponins, in varying amounts. The use of tannins as a biomordant resulted in similar color retention values. Leaching tests indicated that tannins from the bark had a low ΔE at all concentrations, comparable to those of aluminum/calcium mordants, and enhanced the base color. Leaf-derived tannins exhibited very low to low ΔE values, demonstrating greater durability in preventing color leaching. Compared with conventional metal-based materials, tannins provide a viable natural alternative with comparable color stability and eco-friendliness. Ecoprinting applications demonstrated that different leaf combinations influenced the color intensity, with *C. siamea* producing the strongest color when bark tannins were used and *S. mahagoni* producing the strongest color when leaf tannins were used. In addition, the use of tannins did not significantly alter the texture of the fabric, thereby ensuring its suitability for textile applications. In addition to its technical benefits, this approach offers environmental advantages such as reduced chemical waste, water savings, and lower carbon emissions than synthetic mordants do. Economically, it can support small-scale enterprises in eco-friendly textile production, foster sustainable economic growth, and contribute to the circular economy, particularly in communities surrounding *Eucalyptus* plantations. While these findings highlight the potential of *E. urophylla* tannins in natural dyeing, further research is needed to optimize their extraction efficiency, explore their application to different fabric types, and assess their long-term color stability under real-world conditions. In conclusion, tannins from *E. urophylla* bark and leaves are promising biomordants for sustainable ecoprinting, offering an eco-friendly alternative to synthetic mordants while strengthening circular economy initiatives in the textile industry.

1 Introduction

The genus *Eucalyptus* is among the most promising trees for producing wood for forest plantations and is further utilized as a raw material for the pulp and paper industries in several countries, such as Vietnam (Wu et al. 2006), China (Hung et al. 2015), and Indonesia (Wahyudi et al. 2015). An Indonesian pulp company has been growing *Eucalyptus* trees since they are thought to be fast-growing and water-efficient, requiring less water (evapotranspiration) (Alfian et al. 2019); it is well adapted over a wide range of altitudes, temperatures, and annual precipitation (Singh et al. 2022); and its characteristics can be enhanced by hybridization among the species (Prasetyo et al. 2017). Other biomass obtained from Eucalyptus trees, such as bark and leaves, is underutilized and ends up as trash once its wood is harvested for pulp manufacturing. Therefore, these remnants should be used to assist the circular economy.

A circular economy (CE) is different from a linear or recycling economy (Sariatli 2017). CE is a comprehensive strategy for economic growth to help companies, the community, and the environment shift from a conventional linear economy, which follows a 'take, make, dispose' paradigm of production, to a CE that intentionally promotes regeneration and restorativeness. In a CE, litter can be reused, remanufactured, recycled, recovered, and maintained in economic activity for as long as possible. Attempts have been made to use the residue material in the form of *Eucalyptus* bark as a source of tannins (Pinto et al. 2013; Abdelkhalek et al. 2020; Amari et al. 2021). Numerous studies have summarized the potential applications of tannins extracted from *Eucalyptus* bark across various fields; in the biomedical domain, they have demonstrated anti-inflammatory, anticancer (specifically against human breast cancer), antimutagenic, antimalarial, antitumor, and antibacterial properties; and in the food and cosmetics industries, they have been explored for their antioxidant, anti-inflammatory, antimelanogenesis, and antiallergic effects. Additionally, it exhibits antibacterial activity, including antiviral and antimicrobial properties where tannins have also been utilized in various products, such as adhesives, biopesticides, and fire retardants; in the textile industry, it functions as an active compound and functional agent, contributing to ultraviolet (UV) protection, acting as a mordant, and serving as a natural dye (Ismayati et al. 2024).

A study reported the use of *Eucalyptus* leaves for cotton dyeing (da Silva et al. 2018). From this point of view, *Eucalyptus* leaf extract provides not only color but also functions as an anti-UV and antimicrobial agent for fabrics, even though *Eucalyptus* leaves have generally been discussed as a source of essential oil (Ratnaningsih et al. 2018; Alfian et al. 2019; Purwoko et al. 2023), fragrance, aromatherapy, cosmetic products, hair care, skincare, and other uses, such as sunscreen formulations and preservatives of cosmetic products (Aswandi et al. 2024).

There are at least three species of *Eucalyptus* planted in North Sumatra, namely, *Eucalyptus grandis, E. pellita*, and *E. urophylla,* which are native to Australia, Papua New Guinea, and East Timor, respectively (Prasetyo et al. 2017), where the three species have been hybridized to improve the quality of wood fibers for pulp production (in terms of long fibers and high yield), as well as to improve growth, productivity, and resistance to pests and diseases (Rezende et al. 2014).

Our first attempt to use *Eucalyptus* leaves as functional agents in textiles, as well as natural colorants, via an ecoprinting technique for creating creative/craft products to support a circular economy, is presented in Fig. 1.

In this chapter, only the residue of *E. urophylla* in the form of bark and leaves is discussed in terms of its phytochemical content, particularly that of tannins, and its ability to bind color to fabrics via an ecoprinting technique. In other words, the ability of tannins to act as biomarkers can be evaluated via color quantification. An experiment using a natural dye derived from Secang wood (*Caesalpinia sappan*) and three types of leaves—Mahogany (*Swietenia mahagoni*), Johar (*Cassia siamea*), and Krei Payung (*Filicium decipiens*)—will present not only their traces but also their quantification values. Furthermore, small-scale enterprises will be introduced as a complement to support a circular economy. Therefore, the objective of

(a) (b)

Fig. 1 Three species of *Eucalyptus* leaves, namely, *E. urophylla, E. grandis,* and *E. pellita* (**a**), and the results of creative/crafted products of ecoprinting on fabrics prove that the leaves are capable of dyeing and binding both the color and leaf traces (**b**)

the study is to utilize both bark and leaves of *E. urophylla* as the binding color agent in the ecoprinting technique.

2 Definitions and Terminologies

2.1 Economic Activities

In Fig. 2, a comparison among linear, recycling, and circular economies is presented, highlighting that the waste remains in the first two, whereas there is no waste available in the latter. In other words, there are material cycles or routes that ensure that there is no additional trash during the product-production process.

In the context of *Eucalyptus*, particularly *E. urophylla* as the tree-producing pulp, the explanations are as follows:

Linear Economy
As the *E. urophylla* tree is harvested for pulp production, the tree debarks, and the log is converted into wood particles before being cooked to produce pulp, which can then be used for papermaking. These activities generate residues in plantations, including bark and canopy parts (branches, twigs, and leaves), as well as waste from pulp and paper production, such as black liquor and other chemicals. Figure 3 shows a schematic of the linear economy of *E. urophylla* as a pulp-producing tree.

Recycle Economy
In a recycling economy, residues from *E. urophylla* plantations, such as canopy parts (twigs and leaves) and bark, can be utilized as boiler materials and for the production of eucalyptus oil and tannins. The chemical waste during the pulping process, called "black liquor," can be recycled into "green liquor" or "white liquor."

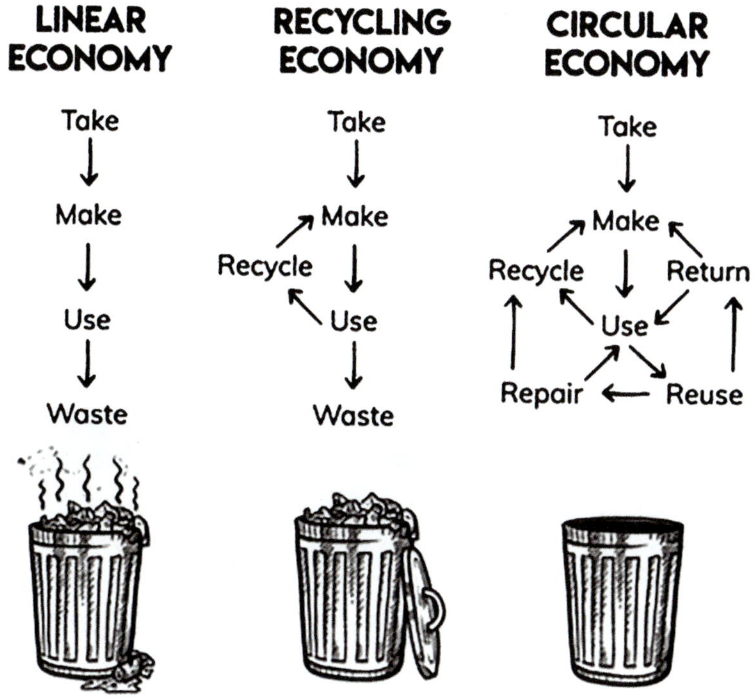

Fig. 2 General comparison among linear, recycling, and circular economies (Stanes 2024)

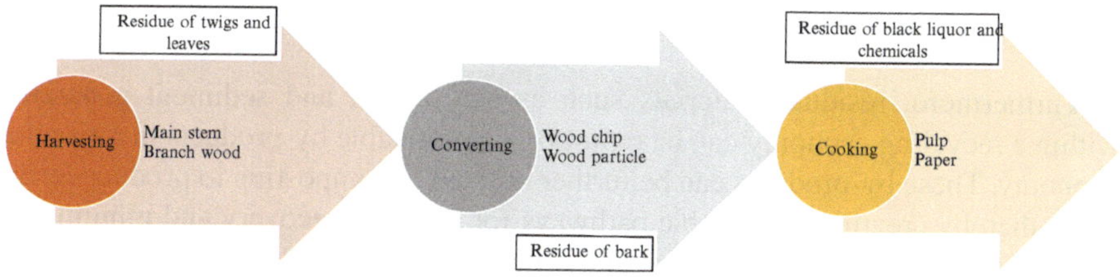

Fig. 3 A scheme of the linear economy of *E. urophylla* when producing pulp/paper resulted in waste both from plantations, such as canopy parts (twigs and leaves), and from industries, such as black liquor and chemicals

Figure 4 shows that the recycling process still generates waste, such as ash (boiler), dregs (after *Eucalyptus* leaf oil extraction), and sediment (after tannin extraction).

Circular Economy

In the case of a circular economy, as depicted in Fig. 5, the residues from *E. urophylla* plantations, such as canopy parts (twigs and leaves) and bark, should be utilized not only for technical products such as boiler materials, eucalyptus oil, and sources of tannins but also for creative products such as natural dyes and ecoprinting products.

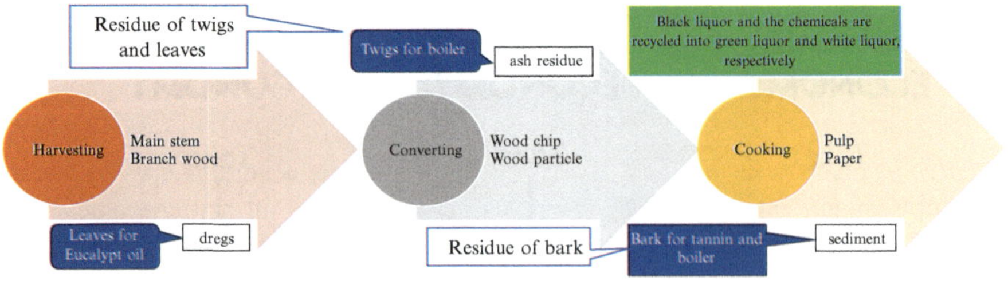

Fig. 4 A scheme of the recycling economy of *E. urophylla* when producing pulp/paper from the main stem or branch wood. Even though residues both from plantations, such as canopy parts (twigs and leaves), and from industries, such as black liquor and chemicals, have been recycled into value-added products, they still generate waste, such as ash, dregs, or sediment

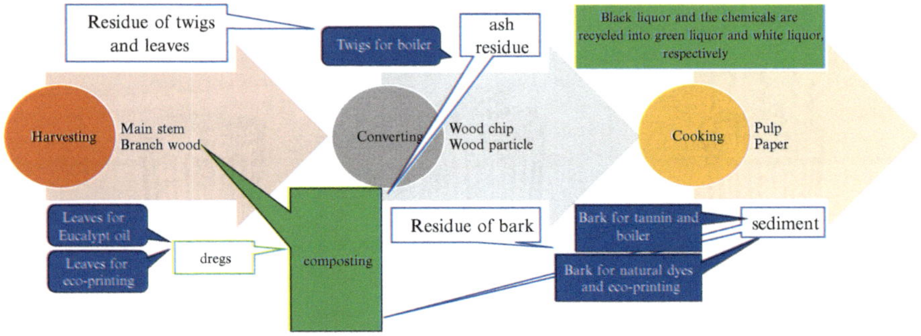

Fig. 5 A scheme of the circular economy of *E. urophylla* when producing pulp/paper from the main stem or branch wood. The diagram presents several routes, illustrating various possibilities for generating value-added products from *E. urophylla* biomass

Furthermore, residual materials such as ash, dregs, and sediment generated within a recycling economy can be repurposed as valuable by-products in a circular economy. These by-products can be further utilized in composting to produce fertilizers, thereby creating sustainable pathways for resource recovery and minimizing waste. This effort was patented by our colleagues who developed fertilizers from *Eucalyptus* waste.

In chemical processing, the pathway aligns with the principles of a recycling economy, as the conversion of "black liquor" to "green liquor" and subsequently to "white liquor" is well established in industrial applications.

2.2 Natural Dyes

Natural dyes are sustainable and eco-friendly alternatives that are nontoxic, renewable, biodegradable, and environmentally friendly. Almost every part of a plant—flowers, fruits, leaves, seeds, bark, stems, and roots—can be used to produce these dyes. They are derived from secondary metabolites produced by plants, microbes, and animals through biosynthetic processes. Although these metabolites play roles in life support, they are not essential for survival. Common secondary metabolites in plants include alkaloids, flavonoids, steroids, saponins, terpenoids, and tannins. Tannins are phenolic chemicals with molecular weights ranging from 500 to 3000.

These compounds are biologically active, play key roles as bioactivators, and act as natural protectors against pests and diseases both within the plant and in its surrounding environment. In this context, natural dyes are used to impart color to fabrics. Our recent study on natural dyes from sikkam (*Bischofia javanica* Blume) was published by Risnasari et al. (2025).

2.3 Ecoprinting Technique

Transferring colors and shapes onto fabrics via direct contact between the fabric and the plant biomass is known as ecoprinting. Color pigments in biomass are released during the heating/steaming process, allowing for color transfer. In contrast to handcrafted batik, ecoprinting involves no wax, requires fewer chemicals, and is easier to apply because the form and pattern of the biomass naturally create traces. Our ecoprinting work has been published in two publications related to business development (Nuryawan et al. 2020) and community engagement (Risnasari et al. 2021). These efforts proved that this activity is not only related to technical aspects but also economically commercial.

3 Key Developments

Origin of Tannin

The leaf and bark residues of *E. urophylla* were provided by PT. Toba Pulp, Lestari Tbk, Sosor Ladang Village, Parmaksian Toba Samosir Regency, North Sumatra, Indonesia. These two residues were extracted to obtain the tannin mixture, which was used as a biomordant. For comparison, metal materials were prepared from aluminum/calcium derived from aluminum sulfate ($Al_2(SO_4)_3$), sodium carbonate (Na_2CO_3), and calcium oxide (CaO). Secang wood extract (*Caesalpinia sappan*) was used as a natural dye base for linen fabrics. Three types of leaves were prepared for ecoprinting—*S. mahagoni*, *C. siamea* (Johar), and *F. decipiens* (Krei Payung). Ferrous sulfate ($FeSO_4$) was applied to the blanket fabrics. The other complementary materials used were soft detergents for scouring and washing linen fabrics before and after the ecoprinting treatment, respectively.

Phytochemical Screening

Before ecoprinting, the leaf and bark extracts of *E. urophylla* were subjected to phytochemical screening to analyze their bioactive compounds. Phytochemical screening involves observing color changes, precipitation, or foam formation in line with the reagents used in both the leaf and bark extracts to identify the presence of secondary plant metabolite compounds. In this study, alkaloids, flavonoids, terpenoids, steroids, saponins, and tannins were examined via test solutions, as previously described (Lestari et al. 2020). Briefly, for the alkaloid test, a small amount of 1% HCl was added to the extracted sample, followed by the addition of 1 mL of Mayer's reagent. The white precipitate indicated the presence of an alkaloid. Wagner's reagent was then added with the same treatment. A brown precipitate was considered positive for alkaloid compounds. To determine the flavonoid content, the extracted sample was mixed with Mg powder and a few drops of concentrated HCl. A pink, magenta, or orange color indicates the presence of flavonoid compounds. To test the terpenoid content, a small amount of anhydrous acetic acid and one drop of H_2SO_4 (Liberman-Buchard reagent) were added to the extracted sample. A brownish-red or brownish-pink ring indicates the presence of a terpenoid compound. For the steroid test, a small amount of anhydrous acetic acid and one drop of Liberman-Buchard reagent were added to the extracted sample. The blue-green color indicates the presence of a steroid compound. For the saponin test, 10 mL of distilled water was added to the extracted sample, which was then shaken vigorously for 30 s. If a stable foam was obtained, it was considered a saponin compound. Finally, the tannin test was performed, where 10 mL of distilled water was added to the extract and boiled. A few drops of $FeCl_3$ were then added to the solution. The presence of a brownish-green or bluish-black color indicated the presence of a tannin.

Process of Extracting Eucalyptus Biomass

The leaves and bark of *E. urophylla* were sun-dried and ground into small pieces. The particles were weighed to determine the dried sample weight after being further dried in an oven at 80 °C for 24 h. They were then extracted via a pressure cooker set to a 1:10 (w/v) (g/ml) particle-water ratio for 30 min at 1000 watts of power. The resulting solution was filtered and evaporated until crystallization was achieved. To determine the crystal weight, the crystals were ground, sieved to a size of 40 mesh, and weighed.

Scouring Process

Scouring is the process of cleaning, exposing pores, and eliminating contaminants or dirt from the fabric surface. The procedure involved soaking the materials for 15 min in a mild detergent solution, followed by thorough washing and rinsing.

Mordanting Process

The purpose of the mordanting process is to increase the ability of the fabric to absorb color. Initially, boiling water was used to dissolve 14 g/L of $Al_2(SO_4)_3$. Six grams of Na_2CO_3 per liter was added to the solution after it had cooled and was gently agitated. After soaking for 12 h, the cloth was washed with fresh water and air-dried at room temperature. The materials were then submerged in a second solution composed of 10 g of CaO/L. After the solution cooled, the cloth was removed, squeezed to extract the solution, and allowed to dry at room temperature. The fabric

(a) (b) (c)

Fig. 6 Ecoprinting steps include (**a**) arranging the leaves, (**b**) covering them with a blanket and plastic, and (**c**) rolling and getting ready to steam

was washed with clean water before being subjected to ecoprinting. The fabric was soaked for 24 h in a tannin solution containing 2%, 6%, 10%, 15%, or 20% of its dry weight. The samples were squeezed and dried.

Ecoprinting Process

Mordant fabrics, which use either metals or biomordants, were soaked in natural dyes of Secang wood extract for 24 h. After squeezing and stretching, the three species of leaves were arranged on the background color of the Secang wood fabric, as depicted in Fig. 6a. The mordanted fabrics with leaves were then covered with linen blankets and plastic (Fig. 6b). Prior to covering, the blanket was dipped for 10 s in an $FeSO_4$ solution (7 g/L). Four layers of mordanted and colored fabrics, leaves, blankets, and plastics were rolled and steamed for 2 h at 100 °C (Fig. 6c). The control was the ecoprinting of fabrics without mordanting. All the experiments were conducted at room temperature with a humidity between 30% and 50%.

Retention Values

The retention values were calculated by weighing the fabric before and after treatment with natural dyes, including mordant treatment, and comparing it with the volume of natural dyes used.

Color Difference Value

Color measurements were carried out via a colorimeter (CHN Spec CS-10) and the CIE Lab method, as defined by the *Commission Internationale de l'Éclairage* (CIE) or *International Commission on Illumination*, which produces L^*, a^*, and b^* values. These values are used to obtain the value of the color difference (ΔE) where ΔL represents the lightness difference (lighter or darker) (L^* sample − L^* control), Δa represents a green or red difference (a^*sample − a^*control), and Δb represents a blue–yellow difference (b^*sample − b^* control), where the value of L represents the color lightness, with a value of 0 for black and 1 for white. The dimension of a is green–red, with $-a^*$ indicating green and $+a^*$ indicating red. The dimensions of "b" are blue–yellow, with $-b^*$ indicating blue and $+b^*$ indicating yellow.

The effect of ΔE was classified according to Table 1.

Table 1 Classification of the color difference before and after the test

Value of ΔE	Effect
<0.2	Not detected/not seen
0.2–1.0	Very low/very little
1.0–3.0	Low/little
3.0–6.0	Moderate
>6	High

Leaching Test

Both Indonesian standards, namely, SNI ISO 105-C06-2010 and SNI ISO 105-B01-2010, were used to evaluate the resistance of the ecoprinted fabrics to detergent washing and ultraviolet sunlight, respectively. The initial test involved washing the fabrics in a washing machine at 40 °C for 45 min, using 5 g per liter (w/v) of *lerak* detergent—a specialized detergent for batik fabrics—in purified water. In the second test, the fabrics were exposed to ultraviolet (UV) light from 9 am to 3 pm.

Data Analysis

The results of the phytochemical screening were tabulated for further analysis and discussion, particularly for tannins derived from the leaves and bark extracts of *E. urophylla*. The experimental data derived from the ecoprinting treatments were then prepared in image form, along with numerical results from the colorimeter examination. Statistics were used to compare the untreated (control) and treated samples (with different concentrations of metal mordant and biomordants or tannins) before and after the leaching test (washing and exposure to UV light).

Brief Discussions Based on Data

Table 2 shows that the bark of *E. urophylla* contains more secondary metabolites than the leaves do. The bark of *E. urophylla* contains flavonoids, saponins, and tannins. Another study stated that the bark of *E. urophylla* contains only flavonoids and tannins (Sartori et al. 2022). However, the leaves contain only tannins.

Since the biomass of *E. urophylla* contains tannins, this experiment was continued because the emphasis of this work was on making biomordants from tannins derived from the bark and leaves. Furthermore, tannins are organic and specific phenolic compounds with complicated structures and additional protein-binding abilities. Phenols found in the hydroxyl group of tannins are essential for forming crosslinks with various dyes and fibers, which aid in color retention (Pisitsak et al. 2018). The total phenol contents of the biomass, bark, and leaves were therefore calculated. Table 3 shows the total phenol content within the biomass of *E. urophylla*.

Although the leaf weight was lower than that of the bark, the total phenol concentration was greater. This tendency was similar to that observed for yield. Table 4 shows that the tannin output from the leaves was greater than that from the bark.

Table 5 shows a visualization of the fabric color via the ecoprinting technique with three leaf patterns—Mahogany, Johar, and Krei Payung leaves. The base color of linen fabrics should be a bright red color (Budiasih et al. 2023) derived from Secang wood or sappanwood. The control, without a mordant, showed unclear

Table 2 Screening of phytochemicals within the *E. urophylla* biomass

Test	Reagent	Biomass of *E. urophylla*	
		Barks	Leaves
Alkaloid	Mayer	(−)	(−)
Flavonoid	Mg powder and HCl	(+)	(−)
Terpenoids	Liberman Buchard	(−)	(−)
Steroid	Liberman Buchard	(−)	(−)
Saponins	Distilled water	(+)	(−)
Tannins	FeCl3	(+)	(+)

Note: (+) positive: contains group compounds; (−) negative: does not contain group compounds

Table 3 Total phenol content within the biomass of *E. urophylla*

Part of the biomass of *E. urophylla*	Weight of the sample (mg)	Absorbance	mg GAE/g sample	mg GAE/100 g sample
Bark	3.047	0.326	0.0823162	8.23
Leaves	1.878	0.515	0.84353621	84.35

Table 4 Yields of tannins derived from the two kinds of biomass

Part of the biomass of *E. urophylla*	Weight of biomass (grams)	Yield (%)
Bark	4290	10.10
Leaves	2200	13.17

dyeing, and only traces of three leaves were observed. In contrast, metal mordants appeared blackish-red in the background, as did traces of leaves, possibly because of the presence of metals. Metal ions interact chemically to form metal coordination bonds. Sappanwood contains brazilin, which creates a natural or purplish-red tint (Kumalasari 2016).

A relatively low tannin concentration resulted in a decreased color. Therefore, increasing the concentration of the tannin mixture improved the color of the mixture. Statistically, only the metal mordant and lower concentration of tannins (2%) resulted in significant differences in the retention of natural dyes between tannins derived from the bark and leaves, as depicted in Fig. 7. Other treatments with relatively high concentrations of tannins (>6%), either from the bark or leaves, presented no significant differences in the retention of natural dyes. Tannins, regardless of their origin (bark or leaves), are capable of binding to natural dyes because of the presence of phenol, thus increasing the retention of natural dyes (Pisitsak et al. 2018).

Figure 8 shows a comparison between the base color and color traces of the three leaves used in ecoprinting with tannins derived from the bark and leaves.

Figure 8 presents the variation values for the base color measurements in all the treatments (blue graph); however, this difference was not statistically significant. This means that the base color of sappanwood may be applied to fabrics as the background color for ecoprinting. With the exceptions of the control and the metal mordant (brown graph) and the metal mordant (gray graph), respectively, the application

Table 5 Visualization of the ecoprinting results without mordant (control), with metal mordant (aluminum/calcium), or with biomordant (tannin) at various concentrations

Treatment	Tannin from bark of *E. urophylla*	Tannin from leaves of *E. urophylla*
Control		
Metal mordant (Aluminum/ Calcium)		
Biomordant Tanin 2%	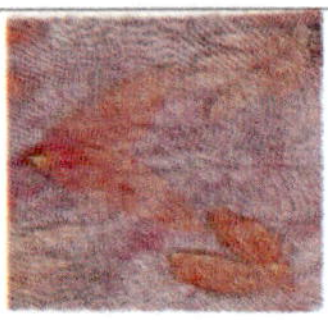	
Biomordant Tanin 6%		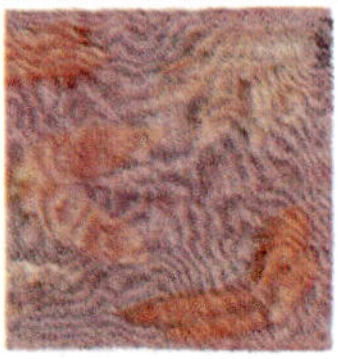
Biomordant Tanin 10%		
Biomordant Tanin 15%		
Biomordant Tanin 20%		

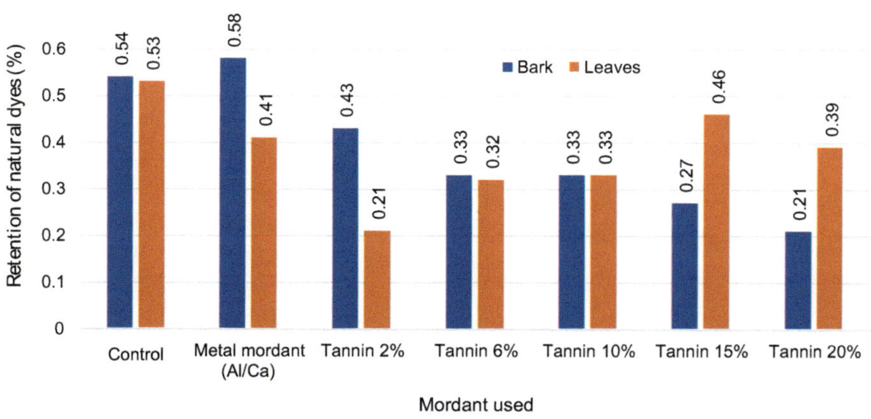

Fig. 7 Comparison of the retention of natural dyes in treated fabrics with different mordants

of the mordant or lack thereof had no effect on color variation when Mahagony and Johar leaves were present. This was not the case for the Krei Payung leaves; even when tannin biomordants were applied, statistically significant differences were detected, particularly for the treatment involving the addition of 6% tannins from the bark (yellow graph). These values were significantly different from those of the control and the samples treated with the metal mixture. From this point on, 6% of the tannins and Krei Payung leaves interacted, resulting in a large color difference ($E > 6$).

Generally, the color difference of the resulting ecoprinting in this work, as presented in Fig. 8, was low/little, moderate, and high, as classified in Table 1, with no values in the range of 0.2–1.0 (not detected until very low). Thus, natural dyes combined with the three leaves were capable of dyeing the fabric with satisfactory results. However, color resistance should be evaluated to test the strength of the dyes bound to the fabrics. Table 6 presents a visualization of the ecoprinted fabrics before and after the leaching test.

After washing and exposure to sunlight (UV), the ecoprinted fabrics were similar and unaltered. However, as presented in Table 7, the statistical analysis reveals differences.

Tannins from the bark and leaves can retain their base color. Johar leaves and tannins from the bark of *E. urophylla*, as well as Mahogany and Johar leaves and tannins from the leaves of *E. urophylla*, were also capable of maintaining the color, pattern, and traces of the ecoprinted fabrics. Compared with those of the control and metal mordant, which varied after the leaching test (the latter did not differ from tannin), the results indicated that without a mordant, the natural colorant would leach and not adhere to the fabrics, whereas some leaching still occurred with a metal mordant. The use of biomordants from tannins from the bark and leaves of *E. urophylla* could preserve natural dyes for the base color of ecoprinting, as well as patterns from the three leaves, in the order Johar > Mahogany > Krei Payung.

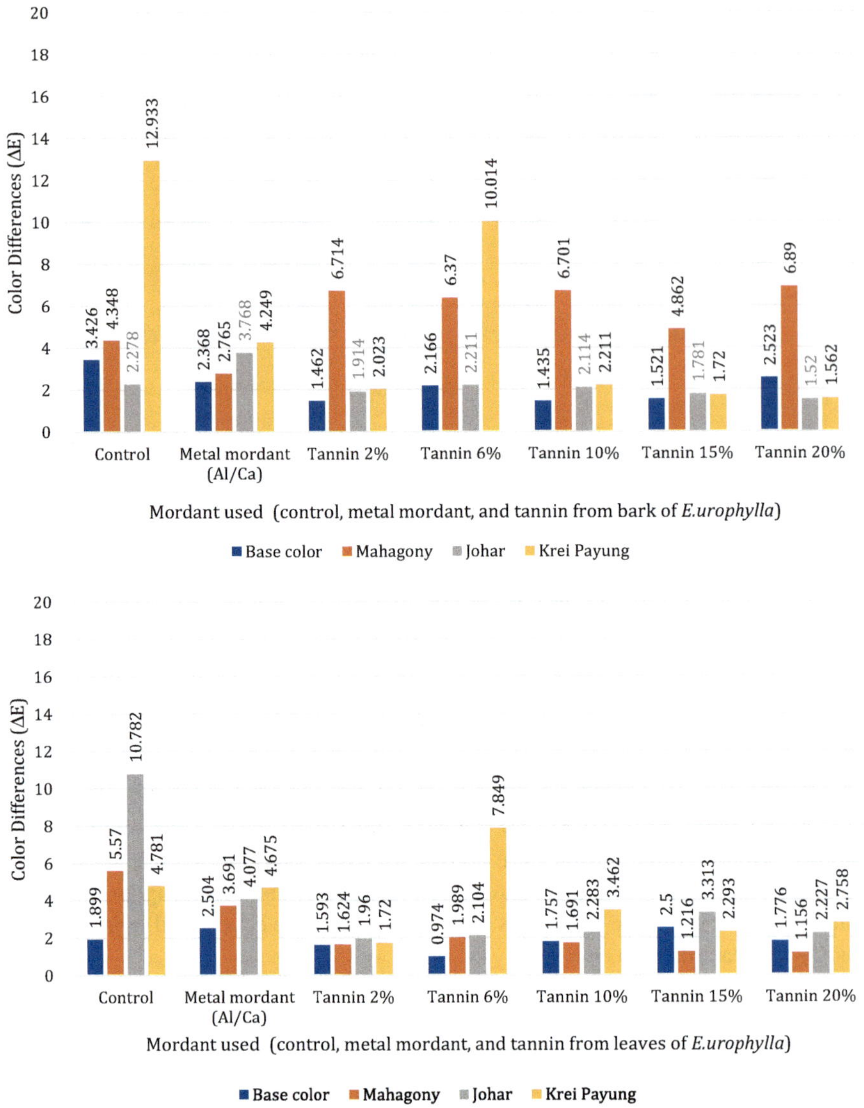

Fig. 8 Comparison of color differences (ΔE) in the ecoprinting of three species of leaves generated from different mordants (without mordants or metal mordants) and from biomordants of tannins derived from the bark (upper) and leaves (lower) of *E. urophylla*

3.1 Conclusions

Tannins from *E. urophylla* bark and leaves are good natural substitutes for traditional metal-based mordants, with similar color stability (which does not change after washing and exposure to UV) and environmental friendliness (which does not

Table 6 Visualization of ecoprinted fabrics from the bark and leaves of *E. urophylla* before and after the leaching test

Treatment	Tannin from the bark of *E. urophylla*		Tannin from leaves of *E. urophylla*	
	Before leaching test	After leaching test	Before leaching test	After leaching test
Control				
Metal mordant (Aluminum/Calcium)				
Biomordant Tanin 2%				
Biomordant Tanin 6%				
Biomordant Tanin 10%				
Biomordant Tanin 15%				
Biomordant Tanin 20%				

leach). Ecoprinting applications demonstrated that different leaf combinations influenced color differences, with the ranking of color stability being Johar > Mahogany > Krei Payung. Additionally, the use of tannins from the bark and leaves of *E. urophylla* did not significantly alter the texture of the fabric, thereby ensuring its suitability for textile applications.

Table 7 Results of the statistical analysis of the differences between the ecoprinted fabrics before and after the leaching test

	Bark				Leaves			
Mordant treatment	Base color	Mahogany	Johar	Krei Payung	Base color	Mahogany	Johar	Krei Payung
Control	a	ab	c	a	a	c	a	ab
Metal mordant (Al/Ca)	a	a	b	a	a	b	a	b
Tannin 2%	a	b	a	a	a	a	a	a
Tannin 6%	a	b	a	b	a	a	a	a
Tannin 10%	a	b	a	a	a	a	a	ab
Tannin 15%	a	ab	a	a	a	a	a	a
Tannin 20%	a	b	a	a	a	a	a	ab

4 Future Perspective

In addition to its technical advantages, the use of biomordants made from tannins—specifically, the bark and leaves of *E. urophylla*—offers environmental benefits such as less chemical waste (converting metal mordants into biomordants), water savings (no leaching), and lower carbon emissions (environmentally friendly) than synthetic mordants do.

Economically, the utilization of biomordants in ecoprinting fabrics presents an opportunity to support small-scale enterprises engaged in sustainable textile production. This approach not only promotes eco-friendly practices but also fosters sustainable economic growth by integrating locally sourced, biodegradable mordants into the value chain. Furthermore, it contributes to the circular economy by minimizing chemical waste, enhancing resource efficiency, and creating livelihood opportunities, particularly for communities residing near Eucalyptus plantations, where biomass-derived mordants can be sustainably harvested and utilized.

5 Concluding Remarks

While these findings highlight the potential of *E. urophylla* tannins in natural dyeing, further research is needed to optimize their extraction efficiency, explore their application to different fabric types, and assess their long-term color stability under real-world conditions. In conclusion, tannins from *E. urophylla* bark and leaves are promising biomordants for sustainable ecoprinting, as this study was conducted, offering an eco-friendly alternative to synthetic mordants while strengthening circular economy initiatives in the textile industry.

References

Abdelkhalek A, Salem MZM, Kordy AM, Salem AZM, Behiry SI (2020) Antiviral, antifungal, and insecticidal activities of Eucalyptus bark extract: HPLC analysis of polyphenolic compounds. Microb Pathog 147:104383

Alfian Z, Marpaung H, Taufik M, Lenny S, Andriayani, Samosir SJ (2019) GC–MS Analysis of chemical contents and physical properties of essential oil of *Eucalyptus grandis* from PT. Toba Pulp Lestari. Asian J Chem 31(10):2319–2322

Amari M, Khimeche K, Hima A, Chebout R, Mezroua A (2021) Synthesis of green adhesive with tannin extracted from Eucalyptus bark for potential use in wood composites. J Renew Mater 9(3):463–475

Aswandi, Kholibrina CR, Kuspradini H (2024) Chapter 11 Phytochemical, essential oils and product applications from Eucalyptus. In: Lee SH, Lum WC, Antov P, Krišt'ák L, Lubis MAR, Fatriasari W (eds) Eucalyptus engineered wood products and other applications. Springer, Singapore, pp 163–183

Budiasih KS, Rohaeti E, Yunita I, Kamari A (2023) Silica modified natural dye of *Caesalpinnia* sappan L. on cotton fabrics. Al Kimiya: J Ilmu Kim Terap 10(1):31–39

da Silva MG, de Barros MASD, de Almeida RTR, Pilau EJ, Pinto E, Soares G, Santos JG (2018) Cleaner production of antimicrobial and anti-UV cotton materials through dyeing with Eucalyptus leaves extract. J Clean Prod 199:807–816

Hung TD, Brawner JT, Mede R, Lee DJ, Southerton S, Thinh HH, Dieters MJ (2015) Estimates of genetic parameters for growth and wood properties in *Eucalyptus pellita* F. Muell. to support tree breeding in Vietnam. Ann For Sci 72:205–217

Ismayati M, Sholihat NN, Sari FP (2024) Chapter 10 Eucalyptus bark tannin for green chemistry agent. In: Lee SH, Lum WC, Antov P, Krišt'ák L, Lubis MAR, Fatriasari W (eds) Eucalyptus engineered wood products and other applications. Springer, Singapore, pp 137–161

Kumalasari V (2016) Potential of ketapang leaves, mahogany leaves and kecombrang flowers as environmentally friendly batik fabric dyeing alternatives. Jukung J Tek Lingkung 2(1):62–70

Lestari YD, Permatasari S, Oktasari A (2020) Antioxidant activity testing of extract kweni peel (*Mangifera odorata* Griff). Indones J Chem Environ 3(2):11–20

Nuryawan A, Risnasari I, Irwansyah A, Pulungan W, Sinaga AS, Ginting FYE, Munthe M (2020) Eco-print on recycle paper and fabric as main products of business development program of intellectual property campus of Universitas Sumatera Utara: eco-print on recycle paper and fabric. J Saintech Transf 3(1):1–11

Pinto PCR, Sousa G, Crispim F, Silvestre AJD, Neto CP (2013) Eucalyptus globulus bark as source of tannin extracts for application in leather industry. ACS Sustain Chem Eng 1(8):950–955

Pisitsak P, Tungsombatvisit N, Singhanu K (2018) Utilization of waste protein from Antarctic krill oil production and natural dye to impart durable UV-properties to cotton textiles. J Clean Prod 174:1215–1223. https://doi.org/10.1016/j.jclepro.2017.11.010

Prasetyo A, Aiso H, Ishiguri F, Wahyudi I, Wijaya IPG, Ohshima J, Yokota S (2017) Variations on growth characteristics and wood properties of three Eucalyptus species planted for pulpwood in Indonesia. Tropics 26(2):59–69

Purwoko A, Azhar I, Damanik SE, Taufik WBNE (2023) Feasibility of utilization of Eucalyptus leaves (*Eucalyptus* spp) from logging waste in industrial plantation forests to produce essential oils. IOP Conf Series: Earth Environ Sci 1241:012106

Ratnaningsih AT, Insusanty E, Azwin. (2018) Yield and quality of essential oil of *Eucalyptus pellita* at various storage time of raw material. Wahana Forestra: Jurnal Kehutanan 13(2):90–98

Rezende GDS, de Resende MDV, De Assisn TF (2014) Eucalyptus breeding for clonal forestry. In: Challenges and opportunities for the world's forests in the 21st century. Springer, Dordrecht. https://doi.org/10.1007/978-94-007-7076-8_16

Risnasari I, Elfiati D, Nuryawan A, Manurung H, Basyuni M, Iswanto AH, Munir E, Slamet B, Susilowati A (2021) Pelatihan pengolahan limbah tanaman mangrove sebagai bahan pewarna alami pada produk ecoprint di Desa Lubuk Kertang Kabupaten Langkat, Sumatera Utara. Sarwahita: Jurnal Pengabdian Kepada Masyarakat 18(1):68–81

Risnasari I, Solihat NN, Rambey R, Purnawati R, Ismayati M, Sari FP, Husna N, Fatriasari W, Nuryawan A, Slamet B, Budiasih KS, Kurnia I, Laila H, Sari DP (2025) Characterization and application of sikkam (*Bischofia javanica* Blume) as a natural dye and antibacterial agent for eco-printing textiles. Waste Biomass Valorization. https://doi.org/10.1007/s12649-025-02958-1

Sariatli F (2017) Linear economy versus circular economy: a comparative and analyzer study for optimization of economy for sustainability. Visegrad J Bioecon Sustain Dev 1:31–34

Sartori CJ, Mota GS, Mori FA, Miranda I, Quilhó T, Pereira H (2022) Bark characterization of a commercial Eucalyptus urophylla hybrid clone in view of its potential use as a biorefinery raw material. Biomass Convers Biorefinery 12:1541–1553

Singh KP, Siregar IZ, Abad JIM, Karlinasari L (2022) Nondestructive modeling using a drilling resistance tool to predict wood basic density of standing trees in a eucalypts plantation in North Sumatra, Indonesia. Biodiversitas 23(12):6218–6226

Stanes E (2024) Understanding options for waste in clothes. Presentation in Universitas Sumatera Utara

Wahyudi I, Ishiguri F, Makino K, Tanabe J, Tan L, Tuhumury A, Iizuka K, Yokota S (2015) Growth characteristics and wood properties of 26-year-old *Eucalyptus alba* planted in Indonesia. Int Wood Prod J 6:84–88

Wu YQ, Hayashi K, Liu Y, Cai Y, Sugimori M (2006) Relationships of anatomical characteristics versus shrinkage and collapse properties in plantation-grown eucalypt wood from China. J Wood Sci 52:187–196

Conclusion: The Future of Eucalyptus Plantation

This book consolidates current research and insights on *Eucalyptus* spp. as an emerging forest plantation species in Malaysia. Despite limited local studies to date, growing interest from policymakers and industry stakeholders reflects its commercial and ecological potential. *Eucalyptus* has become increasingly relevant in Malaysia due to its rapid growth, high biomass yield, and suitability for marginal or underutilized land, making it an ideal candidate for commercial forest plantations. With rising pressure to conserve natural forests and diversify timber supply, *Eucalyptus* offers a sustainable alternative that supports Malaysia's downstream wood-based industries, particularly in pulp, fiberboard, and engineered wood production. Its adaptability to tropical climates, drought resilience, and carbon sequestration capacity also align with national climate goals under the Paris Agreement. Furthermore, its integration into circular economy models, through the utilization of bark, leaves, and wood residues for tannins, biochar, and eco-products, enhances its economic value. National initiatives such as the Forest Plantation Development Program and increasing R&D efforts from local research institutions further strengthen the case for *Eucalyptus* as a strategic crop that supports sustainable forestry, industrial competitiveness, and rural development in Malaysia and the Southeast region. There are ten chapters in the book that cover the spectrum from upstream to downstream of the eucalyptus plantation. Each chapter explores potential, challenges, and improvements that could help the eucalyptus crop have a promising future in forest plantations.

Eucalyptus, a commercially important genus within the Myrtaceae family comprising over 700 species, has gained global recognition for its rapid growth and versatility in timber, pulp, and bioenergy production. In the Malaysian context, its cultivation holds promise for addressing timber supply gaps and supporting industrial diversification. However, conventional propagation through seeds is often hindered by low-germination rates, seed dormancy, and genetic variability, all of which affect the consistency and productivity of plantations. In response, micropropagation has emerged as a scalable and effective biotechnological solution. Using tissue

S. Mohd Razali et al. (eds.), *Sustainable Eucalyptus Forestry in Southeast Asia*, Forestry Sciences 88, https://doi.org/10.1007/978-3-032-21343-3

culture methods, it allows for the creation of seedlings that are genetically the same, free from diseases, and have better qualities—helping to overcome major challenges in developing large eucalyptus plantations and supporting more productive and sustainable forestry practices.

Soil degradation is driven by erosion, nutrient loss, compaction, and climate variability, which poses a serious threat to the sustainability of agriculture and forest plantations. In this context, innovation-driven agroforestry systems that integrate trees with crops and digital technologies offer a regenerative solution. The application of tools such as geographic information systems (GISs), remote sensing, the Internet of Things (IoTs), and artificial intelligence (AI) allows for precise monitoring, adaptive land-use planning, and real-time soil health management. These smart agroforestry approaches are particularly relevant to eucalyptus-based landscapes, where maintaining soil fertility and moisture is critical for plantation productivity. By enhancing nutrient cycling, microbial diversity, and climate resilience, tech-enabled agroforestry supports sustainable land stewardship while also offering scalable opportunities for smallholder integration and rural development.

The establishment of forest plantations, particularly those dominated by *Eucalyptus* spp., brings about notable changes in land use and vegetation structure, which in turn can significantly alter local microclimatic conditions. These shifts, which include reduced humidity, increased surface temperatures, and changes in wind patterns, may influence regional rainfall dynamics and heighten the risk of forest fires. Eucalyptus plantations are especially vulnerable due to their flammable oils, dense understory, and uniform age structure, which can facilitate rapid fire spread. This volume reviews current research linking eucalyptus plantations with microclimate modification and fire occurrence, highlighting both the extent of scientific understanding and existing gaps. The findings emphasize the importance of incorporating fire risk assessments and climate-adaptive design into plantation planning. Future research should focus on long-term microclimate monitoring, predictive fire modelling, and management strategies that reduce flammability while enhancing ecosystem resilience.

Over the past 15 years, Malaysia has undertaken large-scale planting of both native and non-native tree species, including *Eucalyptus*, as part of its strategy to enhance timber supply and reduce dependency on natural forests. While these forest plantations serve important economic functions—both domestically and for export, they inevitably alter habitat structure and floristic composition, which can influence local biodiversity. Evidence from international studies, including over 50 published works on eucalyptus plantations, reveals that while some vertebrate and invertebrate species can adapt to these environments, others are excluded due to the lack of ecological features found in natural forests. This variation illustrates the value of designing plantations with biodiversity considerations in mind. Incorporating mixed-species plantings, preserving riparian corridors, and enhancing structural complexity can improve the ecological value of eucalyptus plantations. As Malaysia expands its forest plantation sector, integrating such practices will be vital to ensuring that biodiversity conservation and sustainable development goals are not only compatible but mutually reinforcing.

In addition, recent advancements in near-infrared (NIR) spectroscopy have opened new possibilities for assessing wood quality in eucalyptus plantations. NIR technology is a fast, affordable, and non-damaging way to test wood quality, effectively measuring various wood features like its chemical makeup, structure, physical characteristics, and strength. Its application is particularly advantageous in plantation settings where timely and high-throughput assessments are essential for decision-making. Given the growing interest in producing engineered wood products from *Eucalyptus* spp., NIR spectroscopy offers a promising pathway for establishing efficient and reliable quality control protocols. As silvicultural practices continue to evolve, further research into the integration of NIR with forest management systems could standardize its use as a key analytical tool in plantation forestry and wood-based industries.

Forest fire risk remains a significant concern in *Eucalyptus*-dominated landscapes, where the species' flammability and volatile oils can exacerbate fire intensity. Remote sensing technologies, particularly satellite-based platforms, have emerged as powerful tools for monitoring and managing these risks. The findings presented in this chapter highlight how remote sensing has been effectively used to detect, track, and assess forest fires in eucalyptus plantations, offering critical insights into fire dynamics and landscape vulnerability. Case studies from countries such as Australia and Portugal demonstrate the utility of these technologies in both early warning systems and post-fire evaluation. Importantly, such tools not only support ecological monitoring but also enhance community safety and shape stakeholder perceptions and preparedness. In Malaysia, using satellite-based fire detection along with plantation management systems could greatly improve the ability of eucalyptus forestry operations to handle changing climate conditions.

These technological approaches are equally relevant in the context of other climate-sensitive ecosystems, such as mangrove forests. Mangrove ecosystems provide vital ecological, economic, and social services, including carbon sequestration, coastal protection, and biodiversity support. Drawing on Indonesia's large-scale restoration initiatives, this chapter highlights both the challenges and opportunities of rehabilitating degraded mangrove areas through smart plantation models, community-based approaches, and emerging technologies. Barriers such as limited public awareness, funding constraints, and stakeholder conflicts have hindered restoration success, but innovative policies, such as the establishment of Indonesia's Mangrove Peat Restoration Agency (BRGM), have demonstrated the potential for coordinated, multi-stakeholder implementation. The chapter also explores the ecological interface between mangrove forests and eucalyptus plantations, suggesting that integrated landscape approaches could be adapted for coastal zones in Peninsular Malaysia and Sabah. These models have the potential to restore degraded coastlines, generate carbon credits, support ecotourism, and enhance climate resilience and livelihoods in vulnerable coastal communities.

In the theme of the circular economy, the chapters highlight the economic contribution of the *Eucalyptus* tree. The industrial residues of *Eucalyptus urophylla*, particularly bark and leaves, show strong potential as natural sources of tannins for bio-mordant applications in sustainable textile dyeing through eco-printing. This technique offers an eco-friendly alternative to traditional methods by minimizing chemical use and supporting circular economy practices. The study demonstrated that tannin yields from bark were higher than those from leaves, and both produced comparable color retention and leaching resistance when used with natural dyes like *Caesalpinia sappan*. The results highlight the viability of replacing synthetic mordants with eucalyptus-derived tannins without compromising fabric quality or color stability. Additionally, this approach contributes to waste reduction, supports local economies, and opens new opportunities for eco-textile innovation in eucalyptus-growing regions.

Eucalyptus spp. has been extensively planted not only for timber production but also for environmental protection, particularly in mitigating water and wind erosion. However, the pruning and harvesting processes generate significant amounts of residual biomass, which, if unmanaged, can contribute to waste. This chapter has highlighted the untapped potential of eucalyptus biomass as a source of valuable bioproducts, from essential oils to bio-based chemicals, that align with circular economy principles. By valorizing these by-products, smallholders, farmers, and entrepreneurs can create new income streams while reducing environmental impact. Maximizing the utility of eucalyptus by-products thus offers a dual benefit: minimizing biomass waste and enhancing socioeconomic development, especially in plantation-based communities.

Forest plantations play a vital role in relieving pressure on natural forests, supporting their conservation while ensuring a stable supply of forest resources for Malaysia's current and future needs. Among these, eucalyptus plantations have shown particular promise in contributing to sustainable forest management and improving local socioeconomic conditions. This chapter has examined how eucalyptus cultivation supports rural livelihoods through job creation, income generation, infrastructure development, and enhanced resource availability. The plantation industry drives employment in upstream planting operations and stimulates downstream economic activities. Despite these benefits, knowledge gaps remain regarding long-term livelihood outcomes and income distribution. Addressing these uncertainties is essential for optimizing policy frameworks and enhancing equitable benefits from eucalyptus-based forestry. Overall, the socioeconomic potential of eucalyptus plantations underscores their role in aligning economic growth with environmental sustainability.

To fully realize the potential of *Eucalyptus* spp. as a viable forest plantation species in Malaysia, it is imperative to adopt an integrated approach that considers its ecological, technological, and socioeconomic dimensions. This volume has captured the full plantation spectrum—from upstream innovations in micropropagation and biotechnology that enable the production of high-quality, uniform planting stock to midstream strategies such as smart agroforestry and precision soil conservation supported by tools such as GIS, IoT, and AI. At the same time, ecological factors must not be overlooked. Ensuring biodiversity conservation through adaptive models, such as mixed species planting and habitat corridor integration, is essential for minimizing environmental trade-offs. Research on microclimatic interactions and the species' susceptibility to forest fires reinforces the importance of proactive landscape management and remote sensing technologies. At the end of the process, using NIR to quickly check wood quality and turning eucalyptus biomass into eco-friendly products shows a good fit with circular economy ideas. Notably, the ecological interaction between eucalyptus plantations and mangrove ecosystems signals new opportunities for integrated coastal landscape management. From a socioeconomic standpoint, eucalyptus development has shown potential in enhancing rural livelihoods, generating employment, and contributing to regional economic diversification. However, to unlock its full value, Malaysia must address challenges related to land tenure, environmental safeguards, and value chain coordination through targeted policies, inclusive stakeholder engagement, and sustained investment in innovation.

GPSR Compliance

The European Union's (EU) General Product Safety Regulation (GPSR) is a set of rules that requires consumer products to be safe and our obligations to ensure this.

If you have any concerns about our products, you can contact us on ProductSafety@springernature.com

In case Publisher is established outside the EU, the EU authorized representative is:

Springer Nature Customer Service Center GmbH
Europaplatz 3
69115 Heidelberg, Germany

Batch number: 10171518

Printed by Printforce, the Netherlands